THE PNEUMATICS

OF

HERO OF ALEXANDRIA

FROM THE ORIGINAL GREEK

TRANSLATED FOR AND EDITED BY

BENNET WOODCROFT
PROFESSOR OF MACHINERY IN UNIVERSITY
COLLEGE, LONDON

LONDON
TAYLOR WALTON AND MABERLY
UPPER GOWER STREET AND IVY LANE PATERNOSTER ROW
1851

1. The bent Siphon
2. Concentric or inclosed Siphon
3. Uniform discharge Siphon
4. Siphon which is capable of discharging a greater or less quantity of Liquid with uniformity
5. A vessel for withdrawing Air from a Siphon
6. A vessel for retaining or discharging a Liquid at pleasure
7. A Vessel for discharging Liquids of different temperatures at pleasure
8. A Vessel for discharging Liquids in varying proportions
9. A Water Jet produced by mechanically compressed Air
10. A Valve for a Pump
11. Libations at an Altar produced by Fire
12. A Vessel from which the contents flow when filled to a certain height
13. Two Vessels from which the contents flow, by a Liquid being poured into one only
14. A Bird made to whistle by flowing Water
15. Birds made to sing, and be silent alternately by flowing Water
16. Trumpets sounded by flowing Water
17. Sounds produced on the opening of a Temple Door
18. Drinking Horn from which either Wine or Water will flow
19. A Vessel Containing a Liquid of uniform height, although a Stream flows from it
20. A Vessel which remains full, although Water be drawn from it
21. Sacrificial Vessel which flows only when Money is introduced
22. A Vessel from which a variety of Liquids may be made to flow through one Pipe
23. A Flow of Wine from one Vessel, produced by Water being poured into another
24. A Pipe from which flows Wine and Water in varying proportions
25. A Vessel from which Wine flows in proportion as Water is withdrawn
26. A Vessel from which Wine flows in proportion as Water is poured into another
27. The Fire-Engine
28. An Automaton which drinks at certain times only, on a Liquid being presented to it
29. An Automaton which may be made to drink at any time, on a Liquid being presented to it
30. An Automaton which will drink any quantity that may be presented to it
31. A Wheel in a Temple, which, on being turned, liberates purifying Water
32. A Vessel containing different Wines, any one of which may be liberated by placing a certain Weight in a Cup
33. A self-trimming Lamp

34. A Vessel from which Liquid may be made to flow, on any portion of Water being poured into it

35. A Vessel which will hold a certain quantity of Liquid when the supply is continuous, will only receive a portion of such Liquid if the supply is intermittent

36. A Satyr pouring Water from a Wine-skin into a full Washing-Basin, without making the contents overflow

37. Temple Doors opened by Fire on an Altar

38. Other intermediate means of opening Temple Doors by Fire on an Altar

39. Wine flowing from a Vessel may be arrested on the Introduction of Water, but, when the Supply of Water ceases, the Wine flows again

40. On an Apple being lifted, Hercules shoots a Dragon which then hisses

41. A Vessel from which uniform Quantities only of Liquid can be poured

42. A Water-Jet actuated by compressed Air from the Lungs

43. Notes from a Bird produced at intervals by an intermittent Stream of Water

44. Notes produced from several Birds in succession, by a Stream of Water

45. A Jet of Steam supporting a Sphere

46. The World represented in the Centre of the Universe

47. A Fountain which trickles by the Action of the Sun's Rays

48. A Thyrsus made to whistle by being submerged in Water

49. A Trumpet, in the hands of an Automaton, sounded by compressed Air

50. The Steam-Engine   -

51. A Vessel from which flowing Water may be stopped at pleasure

52. A Drinking-Horn in which a peculiarly formed Siphon is fixed

53. A Vessel in which Water and Air ascend and descend alternately

54. Water driven from the Mouth of a Wine-skin in the Hands of a Satyr, by means of compressed Air

55. A Vessel, out of which Water flows as it is poured in, but if the supply is withheld, Water will not flow again, until the Vessel is half filled; and on the supply being again stopped, it will not then flow until the Vessel is filled.

56. A Cupping-Glass, to which is attached, an Air-exhausted Compartment

57. Description of a Syringe

58. A Vessel from which a Flow of Wine can be stopped, by pouring into it a small Measure of Water

59. A Vessel from which Wine or Water may be made to flow, separately or mixed

60. Libations poured on an Altar, and a Serpent made to hiss, by the Action of Fire

61. Water flowing from a Siphon ceases on surrounding the End of its longer Side with Water

62. A Vessel which emits a Sound when a Liquid is poured from it

63. A Water-Clock, made to govern the quantities of Liquid flowing from a Vessel

64. A Drinking-Horn from which a Mixture of Wine and Water, or pure Water may be made to flow alternately or together, at pleasure

65. A Vessel from which Wine or Water may be made to flow separately or mixed

66. Wine discharged into a Cup in any required quantity

67. A Goblet into which as much Wine flows as is taken out

68. A Shrine over which a Bird may be made to revolve and sing by Worshippers turning a wheel

69. A Siphon fixed in a vessel from which the Discharge shall cease at will

70. Figures made to dance by Fire on an Altar

71. A Lamp in which the Oil can be raised by Water contained within its Stand

72. A Lamp in which the Oil is raised by blowing Air into it

73. A Lamp in which the Oil is raised by Water, as required

74. A Steam-Boiler from which a hot-Air blast, or hot-Air mixed with Steam is blown into the Fire, and from which hot water flows on the introduction of cold

75. A Steam-Boiler from which either a hot Blast may be driven into the fire, a Blackbird made to sing, or a Triton to blow a horn

76. An Altar Organ blown by manual Labour

77. An Altar Organ blown by the agency of a Wind-mill

78. An Automaton, the head of which continues attached to the body, after a knife has entered the neck at one side, passed completely through it, and out at the other; the animal will drink immediately after the operation

Editor's Preface

WHILE the Editor of the present work was engaged in writing an *Analytical History of the Steam-Engine*, it became necessary to consult the ancient mechanicians to ascertain who were the inventors of the several parts composing that machine: the earliest writer on the subject appeared to be Hero of Alexandria; throughout whose work so many of the elementary parts of all Steam-Engines, and those also of most other machines are mentioned, that it was thought a translation of Hero's Pneumatics would be acceptable not only to the Engineer but to the scientific world generally. Although at the commencement of his work, Hero states that he has added his own discoveries to those "handed down by former writers, yet in no instance has he pointed out any thing which originated with himself; nor is there any statement in the text, except the one I have just quoted, which would lead the reader to any other conclusion than that the whole is a compilation from the works of those who at that period of time were styled the "ancient philosophers and mechanicians." Those parts of each vessel or instrument which mechanically perform the operations assigned to them are alike, or nearly so, in the four manuscript and the three printed copies of Hero's works which have been consulted by the Editor; but great diversity of form is given to the vessel in which they are placed. The drawings have been made expressly for this work from the best examples. The seventy-eighth proposition is the only instance in which there is an omission of the illustrative drawing, and this occurs in all the copies; the two drawings which are now supplied to that proposition have been made from the descriptions given in the text. For the Translation of hero from the Greek, the valuable assistance of Mr. J. G. Greenwood, Fellow of University College, London, has been obtained: he is the recently appointed Professor of the Languages and Literature of Greece and Rome, in Owen's College, Manchester. It is confidently hoped that this Translation will be found superior to its predecessors in whatever language; and that it will prove not only generally interesting but practically useful.

Translator's preface

CONCERNING Hero of Alexandria, the author of the treatise here translated, little is known with certainty. When his name and the place of his abode have been given, all that can be positively affirmed is exhausted. We are further told by Hero the younger, who is supposed to have written in the seventh century A.D., that Hero, the author of the "Pneumatics", was a pupil of Ctesibius; - a statement sufficiently probable from the character of his works, and strengthened by an inscription prefixed to another work by Hero on the construction of missiles.

Even the precise period at which Hero lived is a debated point. From his own writings all that can be gathered is that he knew the works of Archimedes, and of Philo the Byzantian, who, again, is known to have been a contemporary of Ctesibius; and, as the earliest mention of him by others is as low down as the fourth century A.D., external evidence, even if it were distinct, would be little trustworthy. Such evidence, however, is vague and scanty. The only direct statement bearing on the date of Hero is the assertion that he was the pupil of Ctesibius. The date of Hero therefore depends on the date of Ctesibius, and this has been variously fixed by different chronologists.

Clinton, (F. H. vol. iii. pp. 535, 538,) who puts Hero as low down as the end of the second century B.C., proceeds on evidence by Athenaeus (vol. iv. p. 174, edit. Schweighäuser) who quotes Aristocles. Now Euergetes II. (Ptolemy VII.) reigned from B.C. 170 to B.C. 117, and hence Clinton assigns Hero, the pupil of Ctesibius, to the reign of Ptolemy VIII, that is, to B.C. 117-81.

Fabricius, on the other hand, (Bibl. Graec, vol. iv. pp. 222, 234, edit. Harl.) setting out from an entirely different datum, places him more than a hundred years earlier, in the time of Ptolemy Philadelphus (Euergetes I.): Athenaeus Mechanicus, (one of the mechanical writers whose works are printed in the *Veterum Mathematicorum Opera*), speaks in a treatise of Ctesibius as a contemporary. This treatise is dedicated to a Marcellus, and Fabricius, assuming, after Hero junior, this Marcellus to be the conqueror of Syracuse, has hence assigned Ctesibius and Hero to the reigns of the second and third Ptolemies (B.C. 285-222).

Of these conflicting dates that assigned by Clinton has been generally adopted. The question is discussed at some length by Schweighäuser, in a note on the passage of Athenaeus referred to above: he deems the identification of the patron of Athenaeus Mechanicus with the conqueror of Syracuse to be unwarranted, and, besides, thinks it most unlikely that at so early a period a Greek should dedicate a work on military engines to any Roman. But from an expression employed by Athenaeus, it may be inferred that his patron was a man of very exalted rank; and the second objection from the alleged improbability that a Greek should dedicate such a work to a Roman at that period will hardly be thought to apply at the period referred to, while the skill displayed by Marcellus in the siege of Syracuse, and the regret expressed by him for the fate of Archimedes, (whether genuine or not,) may well have suggested the dedication to him of a work on military engineering. The assumption of Fabricius, then, is, in itself, not to be too hastily rejected; and it will be seen that it is not so irreconcilable with the statement of Aristocles as has been supposed. Fabricius has carried back the date further than his argument requires or even warrants. Marcellus was killed B.C. 208: Athenaeus might have inscribed his work to him about B.C. 212 or 210; at this period, then, we must suppose Ctesibius to have been known as a philosopher, but he may have lived far into the succeeding century, - possibly even into the reign of Euergetes II. (B.C. 170-117); Hero would thus be placed about B.C. 150, a result by no means inconsistent with the statement of Aristocles, since it is not necessary, with Clinton, to assign the whole of the long reign of Euergetes II. to Ctesibius, and then to put Hero so low down as the reign of Ptolemy VIII.

The treatise on Pneumatics was first published in an Italian translation by Aleotti (Bologna, 1547). In 1575 appeared a Latin version by F. Commandine (Urbino, 1575): this translation, through which the work has been most extensively known, was reprinted at Amsterdam and at Paris. Several other translations were made into Italian, and one into Germany (see Fabricius, iv. p. 235). It was not till the year 1693, and subsequently to the appearance of all the versions named above, that the Greek text was published at Paris in the *Veterum Mathematicorum Opera*. The design of this collection was formed by Thevenot, deputy librarian of the Royal Library in the reign of Louis XIV., and after his death it was carried out by De la Hire. Thevenot's plan was to publish an accurate transcript of the MSS. of the several authors. The inevitable obscurity arising from the numerous corruptions which had crept into the manuscripts was to be remedied by an appendix of notes and a Latin translation. But for the Pneumatics of Hero it seemed sufficient to adopt the already well-known translation of Commandine; and, in consequence, of the eight MSS. of this treatise existing in the Royal Library, that one was chosen which most nearly agreed with the Latin version. This MS. was closely followed, and, as might be expected, the printed text is extremely corrupt: not unfrequently entire clauses are wanting, which, ending with the same word as the clause preceding, seem to have been passed over by the transcriber, whose eye, in returning from his copy to the original, rested on the second instead of the first of the two similar words. These defective passages, which appear to have been conjecturally restored by Commandine, have been supplied in the present translation from MSS. of Hero preserved in the British Museum. These MSS. are described in the appendix, where the most important cases in which the printed text has been supplemented, or otherwise amended, from this source are collected. When any words are included in the translation between brackets, it is to be understood that they appear neither in the text nor in any of the MSS. collated, but have been inserted as necessary to the sense.

The other treatises of Hero are:- 1. On the construction of slings. 2. On the construction of missiles. 3. On automata. These are published in Greek and Latin in the *Vet. Math.* 4. On the method of lifting heavy bodies. This treatise has not yet been edited: it exists only in an Arabic translation. 5. On the "dioptra" or spying-tube: also inedited. It exists in manuscript in the Royal Library at Vienna, and among the MSS. of Hero contained in the Library of the University of Strasburgh. Schweighäuser in his notice of these MSS. (ap. Fabric. iv. p. 226), intimates that this treatise is of much interest, and contains an account of the dioptra "newly invented or improved by Hero himself." Some help might perhaps be derived from it towards the settlement of Hero's date, as the dioptra is mentioned and minutely commented on by Polybius. Several other treatises, entirely lost, are enumerated by Fabricius, iv. p. 236.

A question of great interest presents itself as to the claim of Hero to be considered as the inventor of the several machines and methods described by him. In the introduction of the "Pneumatica" he declares that his purpose is to arrange in order the discoveries of his predecessors, and to add to them his own. The treatise on the construction of missiles is ascribed in some MSS. to Ctesibius, (as in one at Leyden, Fabric. iv. p. 229,). Again, it is singular that neither Pliny nor Vitruvius has any reference to Hero, though Ctesibius and his inventions are repeatedly mentioned. Vitruvius (x. 7) minutely describes a machine for raising water to a great height, which he expressly ascribes to Ctesibius; and in the following chapter he treats, at great length, of the construction of water-organs, yet without any notice of Hero. Both Pliny and Vitruvius expressly name Ctesibius as famous for his skill in the invention of pneumatic and hydraulic instruments. Pliny's words are (vii. 38) "*Laudatus est Ctesibius pneumatica ratione et hydraulicis organis repertis.*" Vitruvius, (x. 7, compare also ix. 8,) after his description of the machine for raising water, says "*Nec tamen haec sola ratio Ctesibii fertur exquisita, sed etiam plures et variis generibus ab eo liquore pressionibus coacto spiritus efferre ab natura mutuatos effectus ostenduntur, uti merularum aquae motu voces, atque engibata, quae bibentia tandem movent sigilla, caeteraque quae delectationibus oculorum et aurium usu sensus eblandiuntur.*" He refers the curious to the commentaries of Ctesibius himself. How well this description of Ctesibius' inventions suits the general character of those preserved by Hero, will be manifest at once. Vitruvius, as Schneider has pointed out,* seems to have had no knowledge of Hero's Pneumatics, as both the forcing-pump and the water-organ differ in several important particulars from those of Hero: he does not even notice the application of the forcing-pump in extinguishing conflagrations. This silence on the part of Vitruvius and Pliny, so remarkable on the supposition that Hero was an original discoverer, is more easily accounted for if we regard him rather as the interpreter of Ctesibius.**

For further details on the life and writings of Hero, the reader is referred to Fabricius, iv. pp. 222-239, Smith's Dictionary of Biography, and Baldi *de Vita Heronis*, in his edition of the Belopoeica.

J.G.G.

Jan. 31, 1851

---

*   On Vitruvius, x. 7. The sections of Hero and the corresponding chapters of Vitruvius are minutely compared by Schneider, Vitruv. vol. iii. pp. 283-330.

**   Baldi arrives at the same conclusion: (p. 74) "*Caeterum haud immerito quispiam dubitaverit quam ob rem Architectus Heronis nostri nomen silentio praeterierit. Nos ideo factum putamus quod ille Ctesibio utpote inventori ea tribuere maluerit quae ab Herone locupletiora et illustriora quam ipse a magistro accepisset evulgata fuere.*"

A TREATISE ON PNEUMATICS.

THE investigation of the properties of Atmospheric Air having been deemed worthy of close attention by the ancient philosophers and mechanists, the former deducing them theoretically, the latter from the action of sensible bodies, we also have thought proper to arrange in order what has been handed down by former writers, and to add thereto our own discoveries: a task from which much advantage will result to those who shall hereafter devote themselves to the study of mathematics. We are further led to write this work from the consideration that it is fitting that the treatment of this subject should correspond with the method given by us in our treatise, in four books, on water-clocks. For, by the union of air, earth, fire and water, and the concurrence of three, or four, elementary principles, various combinations are effected, some of which supply the most pressing wants of human life, while others produce amazement and alarm.

But, before proceeding to our proper subject, we must treat of the vacuum. Some assert that there is absolutely no vacuum; others that, while no continuous vacuum is exhibited in nature, it is to be found distributed in minute portions through air, water, fire and all other substances and this latter opinion, which we will presently demonstrate to be true from sensible phenomena, we adopt. Vessels which seem to most men empty are not empty, as they suppose, but full of air. Now the air, as those who have treated of physics are agreed, is composed of particles minute and light, and for the most part invisible. If, then, we pour water into an apparently empty vessel, air will leave the vessel proportioned in quantity to the water which enters it. This may be seen from the following experiment. Let the vessel which seems to be empty be inverted, and, being carefully kept upright, pressed down into water; the water will not enter it even though it be entirely immersed: so that it is manifest that the air, being matter, and having itself filled all the space in the vessel, does not allow the water to enter. Now, if we bore the bottom of the vessel, the water will enter through the mouth, but the air will escape through the hole. Again, if, before perforating the bottom, we raise the vessel vertically, and turn it up, we shall find the inner surface of the vessel entirely free from moisture, exactly as it was before immersion. Hence it must be assumed that the air is matter. The air when set in motion becomes wind, (for wind is nothing else but air in motion), and if, when the bottom of the vessel has been pierced and the water is entering, we place the hand over the hole, we shall feel the wind escaping from the vessel; and this is nothing else but the air which is being driven out by the water. It is not then to be supposed that there exists in nature a distinct and continuous vacuum, but that it is distributed in small measures through air and liquid and all other bodies. Adamant alone might be thought not to partake of this quality, as it does not admit of fusion or fracture, and, when beaten against anvils or hammers, buries itself in them entire. This peculiarity however is due to its excessive density for the particles of fire, being coarser than the void spaces in the stone, do not pass through them, but only touch the outer surface; consequently, as they do not penetrate into this, as into other substances, no heat results. The particles of the air are in contact with each other, yet they do not fit closely in every part, but void spaces are left between them, as in the sand on the sea shore: the grains of sand must be imagined to correspond to the particles of air, and the air between the grains of sand to the void spaces between the particles of air. Hence, when any force is applied to it, the air is compressed, and, contrary to its nature, falls into the vacant spaces from the pressure exerted on its particles: but when the force is withdrawn, the air returns again to its former position from the elasticity of its particles, as is the case with horn shavings and sponge, which, when compressed and set free again, return to the same position and exhibit the same bulk. Similarly, if from the application of force the particles of air be divided and a vacuum be produced larger than is natural, the particles unite again afterwards; for bodies will have a rapid motion through a vacuum, where there is nothing to obstruct or repel them, until they are in contact. Thus, if a light vessel with a narrow mouth be taken and applied to the lips, and the air be sucked out and discharged, the vessel will be suspended from the lips, the vacuum drawing the flesh towards it that the exhausted space may he filled. It is manifest from this that there was a continuous vacuum in the vessel. The same may be shown by means of the egg-shaped cups used by physicians, which are of glass,* and have narrow mouths. When they wish to fill these with liquid, after sucking out the contained air, they place the finger on the vessel's mouth and invert them into the liquid; then, the finger being withdrawn, the water is drawn up into the exhausted space, though the upward motion is against its nature. Very similar is the operation of cupping-glasses, which, when applied to the body, not only do not fall though of considerable weight but even draw the contiguous matter toward them through the apertures of the body. The explanation is that the fire placed in them consumes and rarefies the air they contain, just as other substances, water, air or earth are consumed and pass over into more subtle substances.

*"Glass working was practised by the ancient Egyptians at a very early period of their national existence. Sir J. G. Wilkinson, in his able work on the Manners and Customs of the ancient Egyptians, has adduced three distinct proofs that the art of Glass working was practised in Egypt before the Exodus of the children of Israel from that land, three thousand five hundred years ago. At Beni Hassan are two paintings representing Glass blowers at work, and from the hieroglyphics accompanying them they are shown to have been executed in the reign of the first Osirtasen at the early date above mentioned. Such was the skill of the Egyptians in glass making, that they successively counterfeited the Amethyst

successively counterfeited the Amethyst and other precious stones worn as ornaments for the person. Winckelmann, a high authority, is of opinion that glass was employed more frequently in ancient than in modern times; it was used by the Egyptians even for coffins; (within 1847 a process was patented in England for making Coffins of Glass) they also employed it not only for drinking vessels but for Mosaic work, the figures of deities, and sacred emblems, in which they attained excellent workmanship, and surprising brilliancy of colour.

"It is certain that the glass houses of Alexandria were celebrated among the ancients for the skill and ingenuity of their workmen; and from thence the Romans, who did not acquire a knowledge of the art till a later period, procured all their glass ware. Most of the large cinerary vases in the British Museum, found in Roman barrows which contained bones and bone-ashes, are, probably, the production of extensive Egyptian or Roman works: they are large, and of excellent form and workmanship: but the Glass is somewhat impure, of a greenish tint, has numerous globules and striae, and is not unlike the modern common crown or sheet glass in quality. We have incidentally mentioned the discovery of Glass at Pompeii. Glass vessels have also been found among the ruins of Herculaneum: and it appears that Glass was used for admitting light to dwellings in Pompeii. Mr. Auldjo, of Noel house, Kensington, who resided several years at Naples, states, that he has seen glass in the window-frames of some of the houses of Pompeii. Mr. Roach Smith has a specimen of ancient flat Glass such as he believes to have been used by the Romans, or their predecessors for windows.- Curiosities of Glass making by Apsley Pallat, London, 1849.

Mr. Layard in his interesting work on Nineveh, 1849, London, in Vol.1, page 342, says: I took the instrument, and, working cautiously myself, was rewarded by the discovery of two small vases, one in alabaster, the other in glass (both in the most perfect preservation) of elegant shape, and admirable workmanship. Each bore the name and title of the Khorsabad King, written in two different ways, as in the inscriptions of Khorsabad."

That something is consumed by the action of fire is manifest from coal-cinders, which, preserving the same bulk as they had before combustion, or nearly so, differ very much in weight. The consumed parts pass away with the smoke into a substance of fire or air or earth: the subtlest parts pass into the highest region where fire is; the parts some-what coarser than these into air, and those coarser still, having been borne with the others a certain space by the current, descend again into the lower regions and mingle with earthy substances. Water also, when consumed by the action of fire, is transformed into air; for the vapour arising from cauldrons placed upon flames is nothing but the evaporation from the liquid passing into air. That fire, then, dissolves and transforms all bodies grosser than itself is evident from the above facts. Again, in the exhalations that rise from the earth the grosser kinds of matter are changed into subtler substances; for dew is sent up from the evaporation of the water contained in the earth by exhalation; and this exhalation is produced by some igneous substance, when the sun is under the earth and warms the ground below, especially if the soil be sulphureous or bituminous, and the ground thus warmed increases the exhalation. The warm springs found in the earth are due to the same cause. The lighter portions of the dew, then, pass into air; the grosser, after being borne upwards for a certain space from the force of the exhalation, when this has cooled at the return of the sun, descend again to the surface.

Winds are produced from excessive exhalation, whereby the air is disturbed and rarefied, and sets in motion the air in immediate contact with it. This movement of the air, however, is not everywhere of uniform velocity: it is more violent in the neighbourhood of the exhalation, where the motion began; fainter at a greater distance from it: just as heavy bodies, when rising, move more rapidly in the lower region where the propelling force is, and more slowly in the higher; and when the force which originally propelled them no longer acts upon them, they return to their natural position, that is, to the surface of the earth. If the propelling force continued to urge them onward with equal velocity, they would never have stopped, but now the force gradually ceases, being as it were expended, and the speed of the motion ceases with it.

Water, again, is transformed into an earthy substance: if we pour water into an earthy and hollow place, after a short time the water disappears, being absorbed by the earthy substance, so that it mingles with, and is actually transformed into, earth. And if any one says that it is not transformed or absorbed by the earth, but is drawn out by heat, either of the sun or some other body, He shall be shewn to be mistaken: for if the same water be put into a vessel of glass, or bronze, or any other solid material, and placed in the sun, for a considerable time it is not diminished except in a very small degree. Water, therefore, is transformed into an earthy substance: indeed, slime and mud are transformations of water into earth.

Moreover, the more subtle substance is transformed into the grosser as in the case of the flame of a lamp dying out for want of oil,-we see it for a time borne upwards and, as it were, striving to reach its proper region, that is, the highest of all above the atmosphere, till, overpowered by the mass of intervening air, it no longer tends to its kindred place, but, as though mixed and interwoven with the particles of air, becomes air itself. The same may be observed with air. For, if a small vessel containing air and carefully closed be placed in water with the mouth uppermost, and then, the vessel being uncovered, the water be allowed to rush in, the air escapes from the vessel; but, being overpowered by the mass of water, it mingles with it again and is transformed so as to become water.

When, therefore, the air in the cupping glasses, being in like manner consumed and rarefied by fire, issues through the pores in the sides of the glass, the space within is exhausted and draws towards it the matter adjacent, of whatever kind it may be. But, if the cupping glass be slightly raised, the air will enter the exhausted space and no more matter will be drawn up.

They, then, who assert that there is absolutely no vacuum may invent many arguments on this subject, and perhaps seem to discourse most plausibly though they offer no tangible proof. If, however, it be shewn by an appeal to sensible phenomena that there is such a thing as a continuous vacuum, but artificially produced; that a vacuum exists also naturally, but scattered in minute portions; and that by compression bodies fill up these scattered vacua, those who bring forward such plausible arguments in this matter will no longer be able to make good their ground.

Provide a spherical vessel, of the thickness of metal plate so as not to be easily crushed, containing about 8 cotylae (2 quarts). When this has been tightly closed on every side, pierce a hole in it, and insert a siphon, or slender tube, of bronze, so as not to touch the part diametrically opposite to the point of perforation, that a passage may be left for water. The other end of the siphon must project about 3 fingers' breadth (2 in.) above the globe, and the circumference of the aperture through which the siphon is inserted must be closed with tin applied both to the siphon and to the outer surface of the globe, so that when it is desired to breathe through the siphon no air may possibly escape from the vessel. Let us watch the result. The globe, like other vessels commonly said to be empty, contains air, and as this air fills all the space within it and presses uniformly against the inner surface of the vessel, if there is no vacuum, as some suppose, we can neither introduce water nor more air, unless the air contained before make way for it; and if by the application of force we make the attempt, the vessel, being full, will burst sooner than admit it. For the particles of air cannot be condensed, as there must in that case be interstices between them, by compression into which their bulk may become less; but this is not credible if there is no vacuum nor again, as the particles press against one another throughout their whole surface and likewise against the sides of the vessel, can they be pushed away so as to make room if there is no vacuum. Thus in no way can anything from without be introduced into the globe unless some portion of the previously contained air escape; if, that is to say, the whole space is closely and uniformly filled, as the objectors suppose. And yet, if any one, inserting the siphon in his mouth, shall blow into the globe, he will introduce much wind without any of the previously contained air giving way. And, this being the uniform result, it is clearly shewn that a condensation takes place of the particles contained in the globe into the interspersed vacua. The condensation however is effected artificially by the forcible introduction of air. Now if, after blowing into the vessel, we bring the hand close to the mouth, and quickly cover the siphon with the finger, the air remains the whole time pent up in the globe; and on the removal of the finger the introduced air will rush out again with a loud noise, being thrust out, as we stated, by the expansion of the original air which takes place from its elasticity. Again, if we draw out the air in the globe by suction through the siphon, it will follow abundantly, though no other substance take its place in the vessel, as has been said in the case of the egg. By this experiment it is completely proved that an accumulation of vacuum goes on in the globe; for the particles of air left behind cannot grow larger in the interval so as to occupy the space left by the particles driven out. For if they increase in magnitude when no foreign substance can be added, it must be supposed that this increase arises from expansion, which is equivalent to a re-arrangement of the particles through the production of a vacuum. But it is maintained that there is no vacuum; the particles therefore will not become larger, for it is not possible to imagine for them any other mode of increase. It is clear, then, from what has been said that certain void spaces are interspersed between the particles of the air, into which, when force is applied, they fall contrary to their natural action.

The air contained in the vessel inverted in water does not undergo much compression, for the compressing force is not considerable, seeing that water, in its own nature, possesses neither weight nor power of excessive pressure. Whence it is that, though divers to the bottom of the sea support an immense weight of water on their backs, respiration is not compelled by the water, though the air contained in their nostrils is extremely little. It is worth while here to examine what reason is given why those who dive deep, supporting on their backs an immense weight of water, are not crushed. Some say that it is because water is of uniform weight: but these give no reason why divers are not crushed by the water above. The true reason may be shewn as follows. Let us imagine the column of liquid which is directly over the surface of the object under pressure, (in immediate contact with which the water is,) to be a body of the same weight and form as the superincumbent liquid, and that this is so placed in the water that its under surface coincides with the surface of the body pressed, resting upon it in the same manner as the previously superincumbent liquid, with which it exactly corresponds. It is clear, then, that this body does not project above the liquid in which it is immersed, and will not sink beneath its surface. For Archimedes has shewn, in his work on 'Floating Bodies,' that bodies of equal weight with any liquid, when immersed in it, will neither project above nor sink beneath its surface: therefore they will not exert pressure on objects beneath. Again, such a body, if all objects which exert pressure from above be removed, remains in the same place; how then can a body which has no tendency downward exert pressure? Similarly, the liquid displaced by the body will not exert pressure on objects beneath; for, as regards rest and motion, the body in question does [not] differ from the liquid which occupies the same space.

Again, that void spaces exist may be seen from the following considerations: for, if there were not such spaces, neither light, nor heat nor any other material force could penetrate through water, or air, or any body whatever. How could the rays of the sun, for example, penetrate through water to the bottom of the vessel? If there were no pores in the fluid, and the rays thrust the water aside by force, the consequence would be that full vessels would overflow, which however does not take place. Again, if the rays thrust the water aside by force, it would not be found that some were reflected while others penetrated below; but now all those rays that impinge upon the particles of the water are driven back, as it were, and reflected, while those that come in contact with the void spaces, meeting with but few particles, penetrate to the bottom of the vessel. It is clear, too, that void spaces exist in water from this, that, when wine is poured into water, it is seen to spread itself through every part of the water, which it would not do if there were no vacua in the water. Again, one light traverses another; for, when several lamps are lighted, all objects are brilliantly illuminated, the rays passing in every direction through each other. And indeed it is possible to penetrate through bronze, iron, and all other bodies, as is seen in the instance of the marine torpedo.

That a continuous vacuum can be artificially produced has been shown by the application of a light vessel to the mouth, and by the egg of physicians. With regard, then, to the nature of the vacuum, though other proofs exist, we deem those that have been given, and which are founded on sensible phenomena, to be sufficient. It may, therefore, be affirmed in this matter that every body is composed of minute particles, between which are empty spaces less than the particles of the body, (so that we erroneously say that there is no vacuum except by the application of force, and that every place is full either of air, or water, or some other substance), and, in proportion as any one of these particles recedes, some other follows it and fills the vacant space: that there is no continuous vacuum except by the application of some force: and again, that the absolute vacuum is never found, but is produced artificially.

These things having been clearly explained, let us treat of the theorems resulting from the combination of these principles; for, by means of them, many curious and astonishing kinds of motion may be discovered. After these preliminary considerations we will begin by treating of the bent siphon, which is most useful in many ways in Pneumatics.

1. The bent Siphon

LET A B C, (fig. 1), be a bent siphon, or tube, of which the leg A B is plunged into a vessel D E containing water.

If the surface of the water is in F G, the leg of the siphon, A B, will be filled with water as high as the surface, that is, up to H, the portion H B C remaining full of air. If, then, we draw off the air by suction through the aperture C, the liquid also will follow from the impossibility, explained above, of a continuous vacuum. And, if the aperture C be level with the surface of the water, the siphon C, though full, will not discharge the water, but will remain full: so that, although it is contrary to nature for water to rise, it has risen so as to fill the tube A B C; and the water will remain in equilibrium, like the beams of a balance, the portion H B being raised on high, and the portion B C suspended. But if the outer mouth of the siphon be lower than the surface F G, as at K, the water flows out; for the liquid in K B, being heavier, overpowers and draws toward it the liquid in B H. The discharge, however, continues only until the surface of the water is on a level with the mouth K, when, for the same reason as before, the efflux ceases. But if the outer mouth of the tube be lower than K, as at L, the discharge continues until the surface of the water reaches the mouth A. If then we wish all the water in the vessel to be drawn out, we must depress the siphon so far that the mouth A may reach the bottom of the vessel, leaving only a passage for the water.

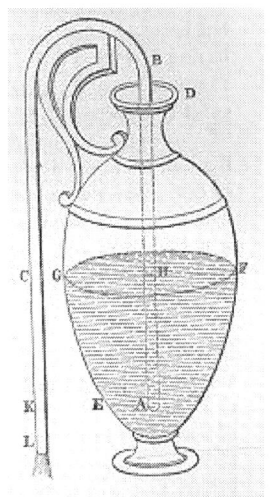

Now some writers have given the above explanation of the action of the siphon, saying that the longer leg, holding more, attracts the shorter. But that such an explanation is incorrect, and that he who believes so would be greatly mistaken if he were to attempt to raise water from a lower level, we may prove as follows. Let there be a siphon with its inner leg longer and narrow, and the outer much less in length but broader so as to contain more water than the longer leg. Then, having first filled the siphon with water, plunge the longer leg into a vessel of water or a well. Now, if we allow the water to flow, the outer leg, containing more than the inner, should draw the water out of the longer leg, which will at the same time draw up the water in the well; and the discharge having begun will exhaust all the water or continue for ever, since the liquid without is more than that within. But this is not found to be the case; and therefore the alleged cause is not the true one. Let us then examine into the natural cause. The surface of every liquid body, when at rest, is spherical and concentric with that of the earth; and, if the liquid be not at rest, it moves until it attains such a surface. If then we take two vessels and pour water into each, and, after filling the siphon and closing its extremities with the fingers, insert one leg into one vessel plunging it beneath the water, and the other into the other, all the water will be continuous, for each of the liquids in the vessels communicates with that in the siphon. If, then, the surfaces of the liquids in the vessels were at the same level before, they will both remain at rest when the siphon is plunged in. But if they were not, as soon as the water is continuous it must inevitably flow into the lower vessel through the channel of communication, until either all the water in both vessels stands at the same height, or one of the vessels is emptied. Suppose that the liquids stand at the same height; they will of course be at rest, so that the liquid in the siphon will also be at rest. If, then, the siphon be conceived to be intersected by a plane in the surface of the liquids in the vessels, even now the liquid in the siphon will be at rest, and, if raised without being inclined to either side, it will again be at rest, and that, whether the siphon is of equal breadth throughout or one leg is much larger than the other. For the reason why the liquid remained at rest did not lie in this, but in the fact that the apertures of the siphon were at the same level. The question now arises why, when the siphon is raised, the water is not borne down by its own weight, having beneath it air which is lighter than itself. The answer is that a continuous void cannot exist; so that, if the water is to descend, we must first fill the upper part of the siphon, into which no air can possibly force its way. But if we pierce a hole in the upper part of the siphon, the water will immediately be rent in sunder the air having found a passage. Before the hole is bored, the liquid in the siphon, resting on the air beneath, tends to drive it away, but the air having no means of escape does not allow the water to pass out: when however the air has obtained a passage through the hole, being unable to sustain the pressure of the water, it escapes.

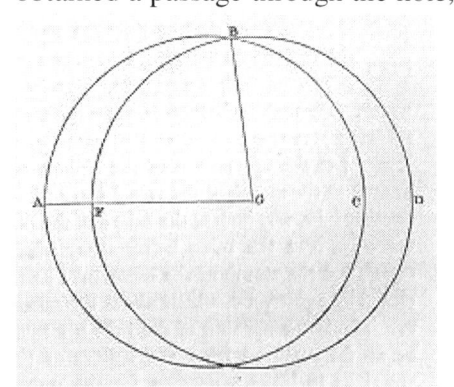

It is from the same cause that, by means of a siphon, we can suck wine upwards, though this is contrary to the nature of a liquid; for, when we have received into the body the air which was in the siphon, we become fuller than before, and a pressure is exerted on the air continuous to us, and this in turn presses on the atmosphere at large until a void has been produced at the surface of the wine, and then the wine undergoing pressure itself will pass into the exhausted space of the siphon; for there is no other place into which it can escape from the pressure. It is from this cause that its unnatural upward movement arises. That the water in the siphon will rest when its surface is spherical and concentric with that of the earth may be shewn otherwise. It is required to prove that a liquid is stationary when its surface is spherical and concentric with that of the earth. If possible let it not be stationary; it will of course become so after being moved; let it then have become stationary. Its surface will now be spherical and concentric with that of the earth, and it will cut the former surface, for, when the same liquid has taken two positions, there must be a line of intersection common to both. Let both surfaces be cut by a plane passing through the centre of the earth; the intersections will be the circumferences of circles concentric with the earth. Let these circumferences be A B C and F B D, (fig. 1 a.). Join B G; B G is equal to each of the lines G F, G A, which is absurd. The liquid will therefore be in equilibrium

2. Concentric or inclosed Siphon

THERE is another kind of siphon called the concentric or inclosed diabetes, the principle of which is the same as that of the bent siphon. As before, let there be a vessel, A B (fig. 2), containing water. Through its bottom insert a tube, C D, soldered into the bottom and projecting below. Let the aperture c of the siphon approach to the mouth of the vessel A B, and let another tube, E F, inclose the tube C D the distance between the tubes being every where equal, and the mouth of the outer tube being closed by a plate, E G, a little above the mouth C. The lower opening of the tube E F must be so far removed from the bottom of the vessel as to leave a passage for the water. These arrangements being completed, if we exhaust, by suction through the mouth D, the air in the tube C D, we shall draw into it the water in the vessel A B, so that it will flow out through the projection of the siphon until the water is exhausted. For the air contained between the liquid and the tube E F, being but little, can pass into the tube C D, and the water can then be drawn after it. And the water will not cease flowing because of the projection of the siphon below -if, indeed, the tube E F were removed, the discharge would cease on the surface of the water arriving at C, in spite of the projection below; but when E F is entirely immersed no air can enter the siphon in place of that drawn off, since the air which enters the vessel takes the place of the water as it passes out -the discharge then, will not cease, for the whole of the outer aperture of the tube, where the water issues forth, is always lower than the surface of the water in the vessel, and, as one level can never be attained, all the water is drained off, attraction being exerted by the deeper column. If we do not choose to draw out the air in the tube C D by suction, water may be poured into the vessel A B until, when it has risen above C, a discharge begins through C D. In this case, again, all the water in the vessel will be drawn out. This instrument is called, as we said before, the inclosed siphon, or the inclosed diabetes.

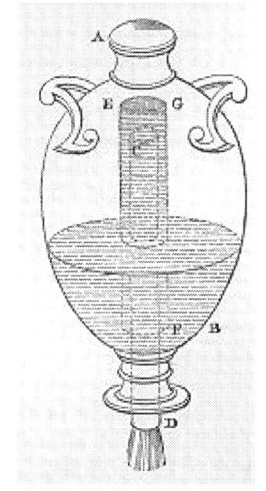

It is evident from what has been proved above that as long as the siphon is stationary the stream through it will be of irregular velocity, for the result is the same as in a discharge through a hole pierced in the bottom of a vessel, where the stream is irregular from the pressure of a greater weight on the discharge at its commencement, and, of a less, as the contents of the vessel are reduced. In like manner, in proportion as the excess of the outer leg of the siphon is greater, the velocity of the stream is greater; for a greater pressure is exerted on the discharge than when the projection of the outer leg below the surface of the water in the vessel is less. Therefore we have said that the discharge through the siphon is always of variable velocity. But we must contrive a siphon in which the velocity of the discharge shall be uniform.

3. Uniform discharge Siphon

LET there be a vessel, A B, (fig. 3.) containing water, on which a small basin, C D, floats, having its mouth covered with the lid C D. Through this lid and the bottom of the basin insert one leg of the siphon soldering it into the holes with tin. Let the other leg be outside the vessel A B, having its mouth lower than the surface of the water in A B. If we draw the air in the siphon through the outer extremity, the water will at once follow because of the impossibility of a continuous vacuum in the siphon; and the siphon, having begun to flow, flows on until it has exhausted all the water in the vessel: but the discharge will be uniform, since the projection of the outer leg below the surface of the water does not vary; for, as the vessel becomes empty, the basin sinks with the siphon. The greater the excess of the outer leg the greater will be the velocity of the discharge, yet still uniform. In the figure, E F G is the siphon described, and the surface of the water is in the line H K.

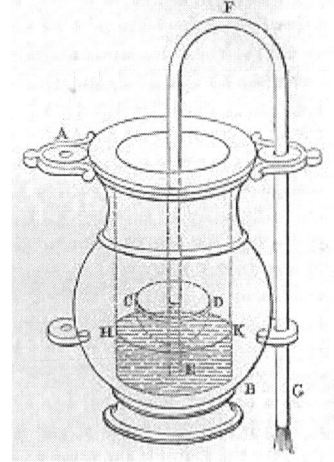

4. Siphon which is capable of discharging a greater or less quantity of Liquid with uniformity

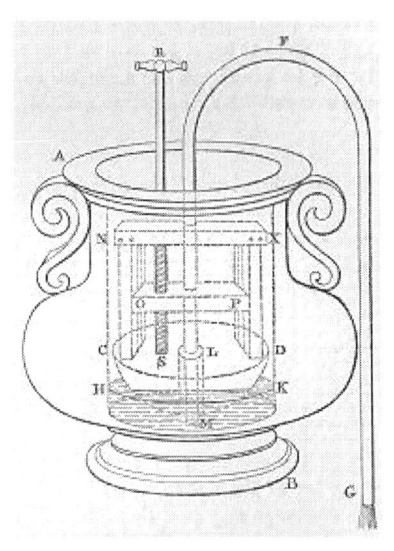

BY the following arrangement we can produce a discharge at once uniform and variable; that is a discharge in which for a certain time at pleasure, the stream continues uniform from the beginning, and again, for any other period, is slower or quicker than before, but still uniform with itself. As before, let A B (fig. 4.) be a vessel of water, and C D a basin. Into the lid and bottom of the basin solder a tube L M wider than the inner leg of the siphon. On the lid place a wooden frame, C N X D, consisting of two upright pieces and a third lying across them on the top. In the inner sides of the upright pieces let grooves be cut down their whole length, along which another piece O P is to move freely. Let R S be a screw, working perpendicularly in the direction of the lid C D, and passing through a hole in O P: in O P let a pin be so fixed as to enter the spiral thread of the screw. The screw must project above N X, and a handle be fastened to its top by which to turn it, and by this means O P can be raised or lowered. Let the inner leg of the siphon be fixed in O P, and pass through the tube L M, so that its mouth may dip into the water in the vessel. Now if, as before, we draw off the liquid through the outer mouth, the siphon will flow with a uniform stream until the whole be exhausted. And when it is wished that a quicker stream should be produced through the siphon, but uniform with itself, let the screw be turned so as to lower the board O P; for then the excess of the outer leg is increased, and thus the stream is still of uniform velocity, but quicker than before. If a still greater velocity is desired, turn the screw so as to lower O P still further; and if a less velocity is sought, let O P be raised. Thus a discharge is produced through a siphon in one sense uniform, in another variable.

5. A vessel for withdrawing Air from a Siphon

TO avoid the necessity of drawing off the water through the mouth, which is only possible in very small siphons, the following contrivance may be used. Take a double tube (fig. 5) one part of which fits into the other, and attach the smaller part to the outer leg of the siphon, so that the discharge may pass through it. Let T N be the smaller tube, and Q U the greater, which must be previously fitted tightly into a vessel, W Y, containing somewhat more water than the siphon will hold, and having an outlet, Z, at the bottom. When it is wished to draw off the water in A B, close the outlet of W Y with the finger, then apply the larger tube Q U to the smaller, and leave the outlet Z free. As the vessel W Y becomes empty the air in the siphon will pass into the exhausted space, and the liquid in A B will follow so as to fill the siphon: then remove the vessel W Y, and let the siphon run.

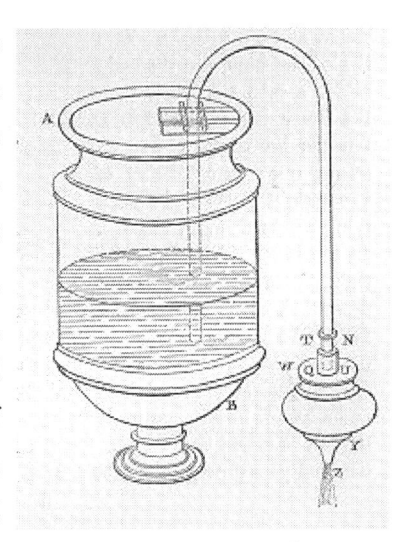

To act properly the siphon must be perpendicular; and this may be secured by fixing two straight bars to the lip of the vessel A B, and placing the inner leg of the siphon between them so as to touch each of the bars: then fasten a small bar crosswise on each side of the leg of the siphon, so as to touch the former bars within. Thus, if the smaller bars touch the larger, the siphon will neither lean sideways nor forwards, but will hang perpendicularly.

6. A vessel for retaining or discharging a Liquid at pleasure

LET us now proceed to construct the necessary instruments, beginning with the less important, as from the elements. The following is a contrivance of use in pouring out wine. A hollow globe of bronze is provided, such as A B (fig. 6) pierced in the lower part with numerous small holes like a sieve. At the top let there be a tube, C D the upper extremity of which is open, communicating with and soldered into the globe. When it is desired to pour out wine, with one hand grasp the tube C D near the mouth C, and plunge the globe into the wine until it is wholly immersed. The wine enters through the holes, and the air within, being driven out, passes through the tube C D: and if, pressing the thumb on the aperture C, you lift the globe out of the wine, the wine contained in the globe will not flow out, as no air can enter to supply the vacuum, for the only entrance is through the mouth C, which is closed by the thumb. When, then, we desire to let the wine flow, we remove the finger, and the air, rushing in, fills the vacuum produced. If we again press the finger on the air-hole C, there will be no discharge until we once more remove the finger from the vent. We may, in like manner, dip the globe into hot or cold water, and then retain or let out the contents at pleasure, until all the water within is exhausted. If the extremity C of the tube C D is bent, the action will be the same, and it is then easier to stop the orifice with the finger.

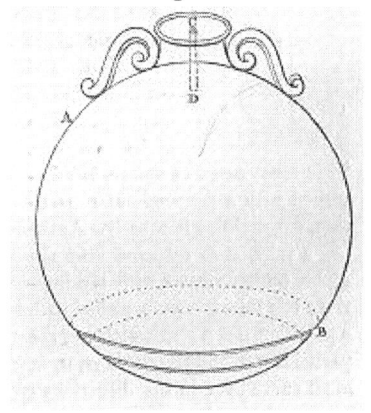

7. A Vessel for discharging Liquids of different temperatures at pleasure

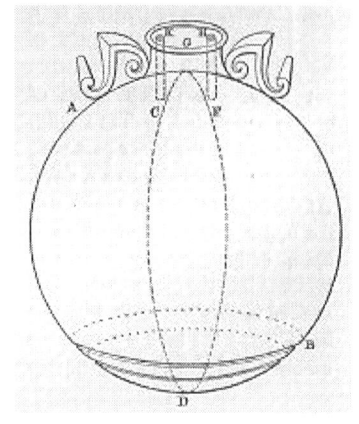

BY the same means it is possible to discharge from the same globe both hot and cold water in any quantity. The globe A B (fig. 7) is constructed as before, but furnished with a perpendicular partition, C D, dividing it equally. At the top a tube, H F, soldered into the globe, communicates with the interior; this tube is also furnished with a partition, C G, a continuation of the partition C D, and its openings, H K, must curve over in the directions C and F. On each side of the partition C D, at the bottom of the globe towards D, let holes be made like those in a cook's ladle. When it is desired to draw hot water, take the apertures H and K by the two fingers and plunge the globe into hot water, and then unclose one of the vents, H, that the air in the hemisphere B C D may be driven out through H: the hot water, entering through the holes, will fill the hemisphere B C D again: stopping the vent H, take the globe out of the water, and its contents will be retained, as the air has no entrance. Then, in like manner, plunge the globe into cold water, unclosing the vent K, and, when the hemisphere A C D is filled, close K and draw the globe out. The globe is now full of hot and cold water, and when it is desired to discharge either of these, unclose the proper vent: in like manner close it again when the discharge is sufficient; and this may be repeated till the contents are exhausted. In the same way it is possible to draw up into and discharge from the same vessel wine, and cold or hot water, and anything else whatever, at any time, and in any quantity, by making the necessary partitions and orifices through which the air may enter into each chamber and leave it again. Instead of the curved outlets, holes may be made in the upper part of the sides of the tube in various directions; and these holes are of course to be closed when it is required to shut out the air. That the holes pierced in the bottom of the vessel may not be seen, both sets may be included in one channel, so that both streams may appear to flow from the same source.

8. A Vessel for discharging Liquids in varying proportions

A JAR can be made receiving and discharging a greater quantity of liquid at one time than at another, and in such a way that, when wine and water are poured into it, it shall discharge at one time pure water, at another time unmixed wine, and, again, a mixture of the two. Its construction is as follows. Let A B (fig. 8) be a pitcher having a partition in the middle, C D. In the partition, near the circumference of the vessel, let small holes be pierced in a curve, as at E. In the opposite side of the partition let there be a circular aperture, F, through which the tube F G H is to be inserted, being soldered into the partition, and reaching nearly to the bottom of the vessel at G. Let the other mouth of the tube H issue at the side of the pitcher, under the handle, and be soldered into the handle which must be hollow, and have a hole on its outer surface at K, which may be closed with the finger when necessary. If, then, closing the vent, as before, we pour any liquid into the jar, the liquid poured into the upper chamber will remain there, not being able to continue its way through the narrow holes into the lower chamber, as there is no other outlet for the air than through the vent K. When, however, we unclose the vent, the liquid will descend into the chamber beneath, and then the jar will hold more. If, then, we first pour in wine so as to fill the chamber B C D, and then, closing the vent, pour water upon it, the two cannot mix, and if we invert the jar it will emit pure water. But, when we unclose the vent, the water continuing to flow, the wine will flow out also, since air can enter through K to fill up the void left; and afterwards the wine will flow out unmixed. We may also pour in the water first, and then, stopping the vent, pour wine upon it, so as to pour out wine for some, wine and water for others, and mere water for those whom we wish to jest with.

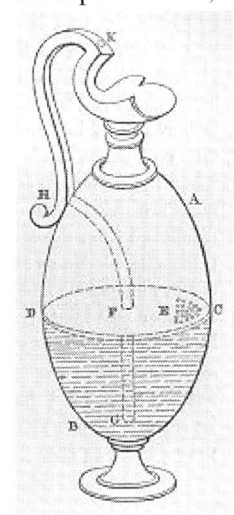

9. A Water Jet produced by mechanically compressed Air

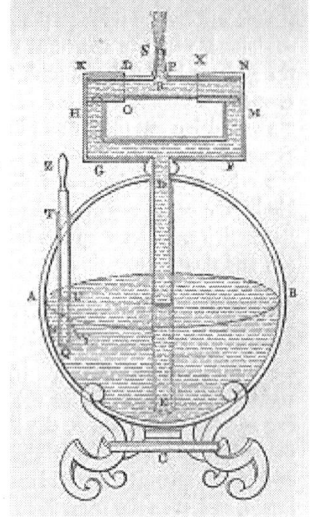

A HOLLOW globe, or other vessel, may be constructed, into which if any liquid be poured, it will be forced aloft spontaneously and with much violence, so as to empty the vessel, though such an upward motion is contrary to nature. The construction is as follows. Let there be a globe, containing about 6 cotylae (3 pints), the sides of which are of metal plate, strong enough to sustain the pressure that will be exerted upon them by the air. Let A B (fig. 9.) be the globe, resting on any base C. Through an aperture in the top of the globe insert a tube, D E, soldered into the globe at the aperture, and projecting a little above it; and reaching to the other extremity, except an interval sufficient for the passage of water. At its upper extremity let the tube D E branch into two tubes, D G and D F, to which two other pipes, G H K L, F M N X, are fastened transversely, communicating with D G, D F. Again, into these transverse pipes, and communicating with them, let another pipe, P O, be fitted, from which a small pipe, R S, projects perpendicularly, communicating with it, and terminating in a small orifice at S. If, then, we take hold of R S and turn round the tube P O, the connection between the corresponding holes will be shut off, so that the liquid which is to be forced up will have no outlet. Now, through another aperture in the globe, let another tube, T U Q, be inserted, closed at the lower extremity Q, and having a hole in the side near the bottom at W. In this hole must be fixed a valve, such as the Romans call assarium, the construction of which we will explain presently. Into the tube T U Q insert another tube, Y Z, fitting tightly. If the tube V Z be drawn out, and water poured into T U Q, it will enter the vessel through the aperture W, (the valve opening into the interior of the vessel), and the air will escape through the pipe O P, which communicates, as we have explained, with the apertures of the pipes G H K L and F M N X. When the globe is half full of liquid, turn the small tube N S so as to break the connection between the corresponding apertures: then depress the tube Y Z and drive out the air and liquid collected in T U Q, which will, on exertion of some force, (as the vessel is full of air and liquid), pass through the valve into the hollow of the globe; and this passage is made possible by the compression of the air into the void spaces dispersed among its particles. Draw up the tube Y Z, in order again to fill T U Q with air, and then, depressing it again, we shall force this air into the globe. By repeating this frequently we shall have a large quantity of air compressed into the globe; for it is clear that the air forced in does not escape again when the rod is drawn up, as the valve, pressed on by the air within, remains closed. if, then, we restore the pipe R S to its up-right position, and re-open the communication between the corresponding apertures at L and X, the liquid will now be forced out, as the condensed air expands to its original bulk and presses on the liquid beneath; and if the quantity of condensed air be large, it will drive out all the liquid, and even the superfluous air will be forced out at the same time.

10. A Valve for a Pump

THE following is the construction of the valve referred to. Take two rectangular plates of bronze of the thickness of a carpenter's rule, and measuring about one finger's breadth (7/10 of an inch) on each side. When these have been accurately fitted to each other, polish their surfaces so that neither air nor liquid may pass between them. Let A B C D, E F G H,(fig. 10) be the plates, and in the centre of one of them, A B C D, bore a circular hole about 1/3 of a finger's breadth (1/4 of an inch) in diameter. Then, applying the side C D to E F, let the plates be attached by means of hinges, so that the polished surfaces may come together. When the valve is to be used, fasten the plate A B C D over the aperture, and any air or liquid forced through will be effectually confined. For by the pressure exerted the hinges move, and the plate E F G H opens readily to admit the air or liquid; which when inclosed in the air-tight vessel, presses on the plate E F G H, and closes the aperture through which the air was forced in.

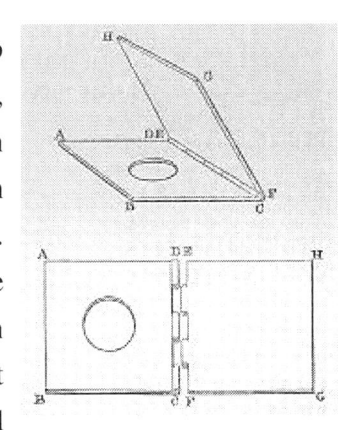

11. Libations at an Altar produced by Fire

TO construct an altar such that, when a fire is raised on it, figures at the side shall offer libations. Let there be a pedestal, A B C D, (fig. 11) on which the figures stand, and also an altar, E F G, perfectly air-tight. The pedestal must also be air-tight, and communicate with the altar at G. Through the pedestal insert the tube H K L, reaching nearly to the bottom at L, and communicating at H with a bowl held by one of the figures. Pour liquid into the pedestal through a hole, M, which must afterwards be closed. Now if a fire be lighted on the altar E F G, the air within it, being rarefied, will descend into the pedestal, and exert pressure on the liquid it contains, which, having no other way of retreat, will pass through the tube H K L into the bowl. Thus the figures will pour a libation, and will not cease so long as the fire remains on the altar. When the fire is extinguished, the libation ceases; and as often as the fire is kindled the same will be repeated. The pipe through which the heat is to pass should be broader towards the middle, for it is requisite that the heat, or rather the vapour from it, passing into a broader space, should expand and act with greater force.

12. A Vessel from which the contents flow when filled to a certain height

THERE are some vessels which emit no stream unless they are filled; but when filled discharge all the liquid they contain. They are made as follows: Let A B C D (fig. 12) be a vessel open at the top, and through its bottom pass a tube, either an inclosed diabetes as E F G, or a bent siphon G H K. When the vessel A B C D is filled, and the water runs over, a discharge will begin through the diabetes, and continue till the vessel is empty, if the interior opening of the diabetes is so near the bottom of the vessel as only to leave a passage for the water.

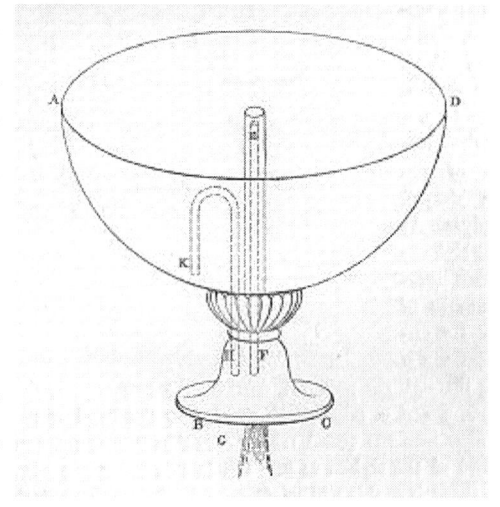

13. Two Vessels from which the contents flow, by a Liquid being poured into one only

IF two vessels, both of them having visible outlets, stand upon a pedestal, and one of them be filled with wine, the other remaining empty, the wine shall not flow out until the empty vessel be filled with water; and then a discharge shall begin, of wine from one, and of water from the other, until both are empty. Such vessels are all called harmonious goblets. Let A B C D (fig. 13) be the pedestal on which the vessels, E and F, stand. In each of them place a bent siphon, G H K in E, and L M N in F, and let the outer extremities of the siphons be shaped like water-pipes. At the bend the siphons must approach nearly to the mouths of the vessels. Let another bent tube, X O P R, passing through the pedestal, connect the two vessels, the extremities of which, X and R, must reach as high as the bend of the siphons. Now pour wine into one vessel, taking care that it does not mount higher than the bend of the siphon at H. Up to this point the wine will not flow out, as there is nothing to originate a discharge through the siphon. But if we pour water into the vessel F, until its surface mounts above the bend of the siphon at M, then the water will descend and pass through the pipe X O P R into the other vessel. Thus a discharge is occasioned of the wine also, and both vessels will continue to run the one with wine, the other with water until both are emptied.

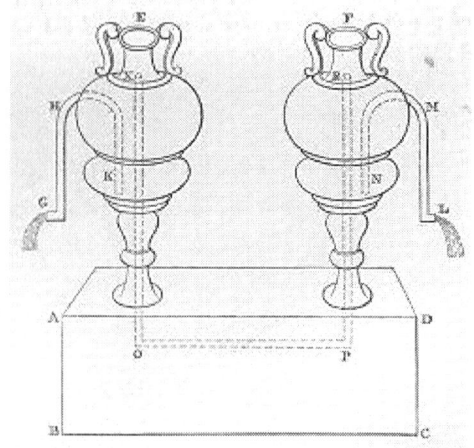

14. A Bird made to whistle by flowing Water

VESSELS may be made such that, when water is poured into them, the note of the black-cap, or a whistling sound, is produced. The following is their construction. Let A B C D (fig. 14) be a hollow air-tight pedestal: through the top, A D, let a funnel, E F, be introduced and soldered into the surface, its tube approaching so near to the bottom as only to leave a passage for the water. Let G H K be a small pipe, such as will emit sound, communicating with the pedestal and likewise soldered into A D. Its extremity, which is curved, must dip into water contained in a small vessel placed near at L. If water be poured in through the funnel E F, the result will be that the air, being driven out, passes through the pipe G H K, and emits a sound. When the extremity of the pipe dips into water a bubbling sound is heard, and the note of the black-cap is produced: if no water is near, there will be a whistling only. These sounds are produced through pipes; but the quality of the sounds will vary as the pipes are more or less fine, or longer, or shorter; and as a larger or smaller portion of the pipe is immersed in the water: so that by this means the distinct notes of many birds can be produced. The figures of several different birds are arranged near a fountain, or in a cave, or in any place where there is running water: near them sits an owl, which, apparently of her own accord, turns at one time towards the birds, and then again away from them; and when the owl looks away the birds sing, when she looks at them they are mute: and this may be repeated frequently.

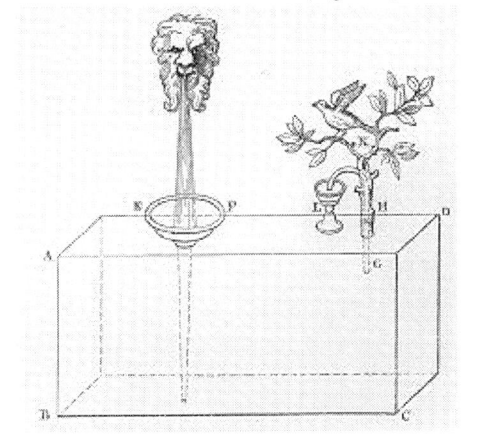

15. Birds made to sing, and be silent alternately by flowing Water

THE construction is after this manner. Let A (fig. 15) be a stream perpetually running. Underneath place an air-tight vessel, B C D E, provided with an inclosed diabetes or bent siphon F G, and having inserted in it a funnel, H K, between the extremity of the tube of which and the bottom of the vessel a passage is left for the water. Let the funnel be provided with several smaller pipes, as described before, at L. It will be found that, while B C D E is being filled with water, the air that is driven out will produce the notes of birds; and as the water is being drawn off N through the Siphon F G after the vessel is filled, the birds will be mute. We are now to describe the contrivance by which the owl is enabled to turn herself towards, or away from, the birds as we have said. Let a rod N X turned in a lathe rest on any support M: round this rod let a tube O P be fitted, so as to move freely about it, and having attached to it the kettle-drum top R S, on which the owl is to be securely fixed. Round the tube O P let a chain pass, the two extremities of which, T U, Q W, wind off in opposite directions, and are attached, by means of two pullies, the one, T U, to a weight suspended at Y, and Q W to an empty vessel Z, which lies beneath the siphon or inclosed diabetes F G. It will be found that while the vessel B C D E is being emptied, the liquid being carried into the vessel Z causes the tube O P to revolve, and the owl with it, so as to face the birds: but when B C D E is exhausted, the vessel Z becomes empty likewise by means of an inclosed or bent siphon contained within it; and then the weight Y, again preponderating, causes the owl to turn away just at the time when the vessel B C D E is being filled again and the notes once more issue from the birds.

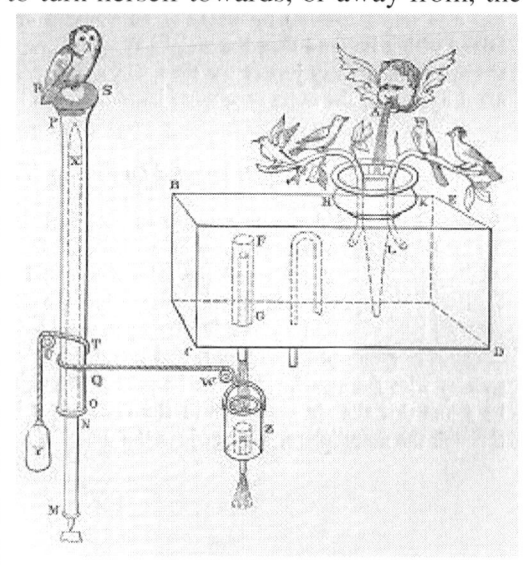

16. Trumpets sounded by flowing Water

IN the same manner as that just described the sound of trumpets can be produced. Insert into a carefully closed vessel the tube of a funnel reaching nearly to the bottom and soldered into the surface of the vessel; and, by its side, a trumpet, provided both with a mouth piece and bell, and communicating at its upper extremity with the vessel. If water be poured through the funnel, it will be found that the air contained in the vessel, as it is being driven out through the mouthpiece, will produce the sound of a trumpet.

17. Sounds produced on the opening of a Temple Door

THE sound of a trumpet may be produced on the opening of the doors of a temple. The following is the construction. Behind the door let there be a vessel, A B C D (fig. 17), containing water. In this invert a narrow-necked vessel, shaped like an extinguisher, F, with which, at its lower extremity, let a trumpet, H K, communicate, provided with bell and mouthpiece. Parallel with the tube of the trumpet, and attached to it, let the rod L M run, fastened, at the lower end, to the vessel F, and having at the other extremity a loop, M: through this loop let the beam N X pass, thus supporting the vessel F, at a sufficient height above the water. The beam N X must turn on the pivot O, and a chain or cord, attached to the extremity X, be fastened, by means of the pulley, P, to the hinder part of the door. When the door is opened, the cord will be stretched, and draw upwards the extremity X of the beam, so that the beam N X no longer supports the loop M; and when the loop changes its position in consequence, the vessel F will descend into the water, and give forth the sound of a trumpet by the expulsion of the air contained in it through the mouthpiece and bell.

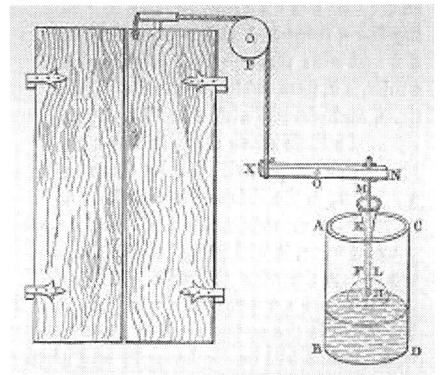

18. Drinking Horn from which either Wine or Water will flow

THERE is a kind of drinking horn, such that if wine be first poured into it, and then water, sometimes the water flows out unmixed, and sometimes the wine. The following is the construction. Let A B C (fig. 18) be a drinking horn, furnished with two partitions, D E and F G: through both of these let a tube, H K, pass, soldered into the partitions, and pierced with a small hole, L situated a little above the partition F G; and under the partition D E let there be a vent, M, in the side of the vessel. If, when these arrangements are complete, we close the passage at C and pour in wine, it will pass through the vent M; and, if we cover M with the finger, the wine in D E F G will be retained. Now, if we pour water into the part A B D E, still closing the vent M, pure water will flow out; but if, while the water is still in the upper part of the vessel, we unclose M, a mixture will be discharged; and when all the water has passed out, the stream will be of pure wine. By frequently unclosing M the discharge may be varied: but the better method is first to pour water into the chamber D E G F, and then, closing the vent, to pour wine upon it. The result will be that sometimes pure water flows out, and again, when the siphon is set free, a mixture; presently, on stopping the vent, pure wine. And this can be done as often as we please.

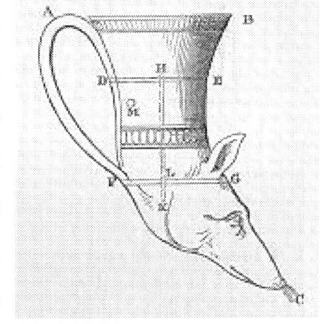

19. A Vessel Containing a Liquid of uniform height, although a Stream flows from it

IF a goblet be placed upon a pedestal, whatever quantity may be drawn from it, it shall always continue full. The construction is as follows. Let A B be a vessel, the mouth of which is closed just at the neck, by the partition C D. Through C D let a tube, E F, be inserted, reaching nearly to the bottom; let another tube, G H, be passed through the bottom of the vessel, reaching nearly up to the partition C D; and in the bottom bore a hole, K, to admit the small tube K L. The vessel A B must stand upon a pedestal, M N O X, through which passes the projection of the tube G H, and another tube S T communicating with the pedestal and the goblet P R. Now let wine be poured through E F into A B (the air will pass out through G H), and, if the tube K L be left open, it will pass through into the pedestal and the goblet P R: but, if K L be closed, the vessel A B will be filled. Let, then, the wine run into the pedestal M N O X and the goblet P R, so that M N O X may be filled as high as the mouth of the tube G H. When this is done, close E, and the wine in A B will no longer flow through K L, for no more air can enter through E to supply the vacuum created. When, therefore, any wine is taken from the goblet, the orifice E must be unclosed, and, the air having found an entrance, the wine will flow again into the pedestal and goblet, until it is full. And this may be done as often as we draw off wine from the goblet. It will be requisite that a small hole be pierced in the side of the pedestal at U, that an equivalent bulk of air may pass into the vessel A B through the orifice G and the hole U.

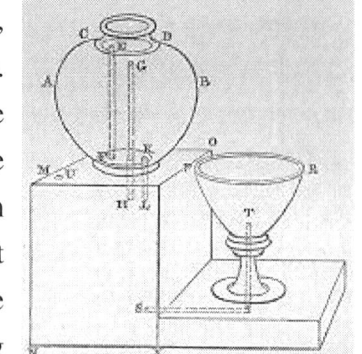

20. A Vessel which remains full, although Water be drawn from it

IF it is desired to adapt this contrivance for use, so that from a goblet occupying any given position a considerable quantity of water may be drawn and yet the goblet remain full, proceed as follows. Let A B (fig. 20) be a vessel containing as much water as will probably be required, and C D a pipe leading from this into a trough beneath, G H. Near the pipe fix a lever beam, E F, and at the extremity E suspend a piece of cork, K, so that it may float in the trough; at the other extremity F let a chain be fastened furnished with a leaden weight, X. Let the whole be so arranged that the cork, floating on the water in G H, closes the mouth of the pipe; yet that, when water has been drawn from the trough, the cork, being heavier than the weight at X, shall sink and open the pipe, so that the water may flow in again and raise the cork. Let L M be the goblet placed in any convenient position, its lip being on a level with the surface of the water in the trough when there is no discharge from the pipe owing

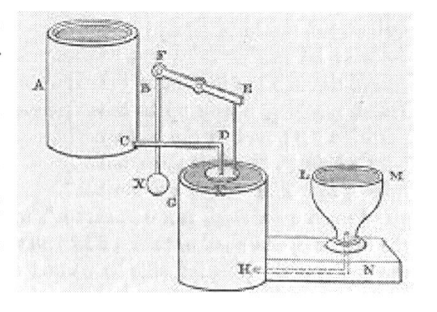

32

to the floating cork: and let the tube H N lead from the trough into the bottom of the goblet. Now if, when the goblet is full, we draw water from it, we shall at the same time reduce the water in the trough; and the cork sinking will unclose the pipe, so that the water, flowing both into the trough and the goblet, will again raise the cork, and the discharge will cease. And this will happen as often as we remove water from the goblet.

21. Sacrificial Vessel which flows only when Money is introduced

IF into certain sacrificial vessels a coin of five drachmas be thrown, water shall flow out and surround them. Let A B C D (fig. 21) be a sacrificial vessel or treasure chest, having an opening in its mouth, A; and in the chest let there be a vessel, F G H K, containing water, and a small box, L, from which a pipe, L M, conducts out of the chest. Near the vessel place a vertical rod, N X, about which turns a lever, O P, widening at O into the plate R parallel to the bottom of the vessel, while at the extremity P is suspended a lid, S, which fits into the box L, so that no water can flow through the tube L M: this lid, however, must be heavier than the plate H, but lighter than the plate and coin combined. When the coin is thrown through the mouth A, it will fall upon the plate H and, preponderating, it will turn the beam O P, and raise the lid of the box so that the water will flow: but if the coin falls off; the lid will descend and close the box so that the discharge ceases.

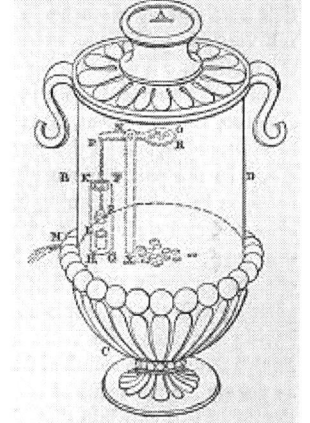

22. A Vessel from which a variety of Liquids may be made to flow through one Pipe

SEVERAL kinds of liquid having been poured into a vessel through one mouth, it is required that through the same pipe they shall flow out separately at pleasure. Let A B (fig.22) be a vessel closed at the neck by the partition C D; and let there be in it several vertical partitions, extending to the partition C D and making as many chambers as we wish to pour in liquids. Suppose, for the present, that these are two in number, and let the partition be E F. In the partition C D pierce fine holes, as in a sieve, opening into each chamber, and air holes, G, H, close to the partition, also opening into the chambers: again, at the bottom let there be small tubes, K, L, communicating with the chambers and opening into the common pipe M. If, having first closed the vents G, H, and the pipe M, we pour one of the liquids through the mouth of the vessel, it will enter into neither chamber, as the air has no means of escape: but if one of the vents be opened, the liquid will pass into that chamber to which the vent belongs; and if, after closing this vent again, we pour in the other liquid and set free the other vent, the liquid will pass into the other chamber. Now let all the vents and the sieve-like holes be closed, and, on opening the pipe M, no discharge can take place until one of the vents be opened; when, the air having found an entrance, the liquid contained in that chamber will flow out. If this vent be closed and the other opened, the same result will follow.

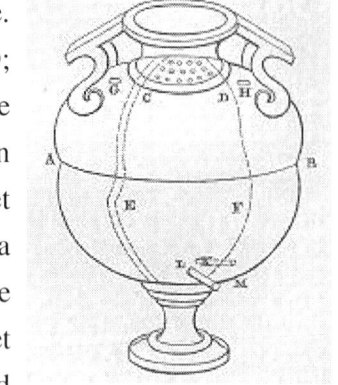

23. A Flow of Wine from one Vessel, produced by Water being poured into another

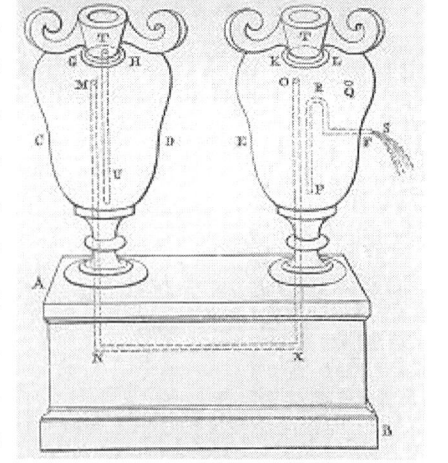

IF of two vessels standing on a pedestal one be full of wine and the other empty, whatever quantity of water be poured into the empty vessel, as much wine shall flow from the other. The following is the construction. On any pedestal, A B (fig. 23) let there be two vessels, C D, E F, having their mouths closed by the partitions G H, K L. Let the tube M N X O pass through the pedestal and bend upwards into the vessels, reaching very nearly to the partitions at M and O. In E F place a bent siphon, P R S, the bend being near the vessel's mouth, and one leg, shaped like a water-pipe, passing outside. Through the partition G H let a funnel, T U, descend almost to the bottom of the vessel, its tube being soldered into the partition. Into the vessel E F pour wine through a hole, Q, which must afterwards be carefully closed again. Now, if we pour water into the vessel C D through the funnel, the contained air will be forced out, and pass through the tube M N X O into E F, and, in its turn, force out the wine contained in E F: and this will happen as often as we pour in the water. It is evident that the air forced out has an equal bulk with the water poured in, and that it will force out as much wine. If no bent siphon be used, but merely a pipe at S, the effect will be the same, unless the force of the water be too great for the pipe.

24. A Pipe from which flows Wine and Water in varying proportions

LET there be an empty vessel, and another containing wine: whatever quantity of water we pour into the empty vessel, the same quantity of wine and water mixed may be drawn off through a pipe in any proportion we please; such, for instance, that there may be two parts of water to one of wine. Let A B (fig 24) be an empty vessel, either a cylinder or a rectangular parallelopiped: by the side of this, and on the same base, place another vessel, C D, perfectly air-tight, and, like A B, either a cylinder or a rectangular parallelopiped; but the base of A B must be twice as great as that of C D, as the water is to be the double of the wine. Near C D place another air-tight vessel, E F, into which the wine is to be poured; and between the vessels C D, E F, let a tube run, G H K, perforating and soldered into their coverings. In E F let there be a bent siphon, L M N, the inner leg of which must reach almost to the bottom of the vessel, leaving only a passage for the water, and the other, being bent within the vessel, lead into the next vessel, O X. From this vessel let the tube P R lead through all the vessels, or be carried under the pedestal on which they stand, that it may readily pass near the bottom of the vessel A B. Let another tube, T S, connect the vessels A B, C D, and near the bottom of A B place a small pipe, U, which with P R must be included in a larger pipe, Q W, provided with a cock by means of which it may be opened or shut at pleasure. When these preparations have been made, close the pipe Q W, and pour water into the vessel A B; a part, viz. one half, will pass into C D, through the tube S T, and the water which falls into C D will force out a mass of air equal to itself through G H K into the vessel E F; in like manner this air will force an equal quantity of wine into the vessel O X through L M N. Now, if we open the pipe Q W, the water poured into the vessel A B and the wine carried out of O X through the tube P R will flow through it together and thus what was proposed will be done. The vessels will be empty again when, the mixed liquid having been all discharged, the air enters them through the tube P R.

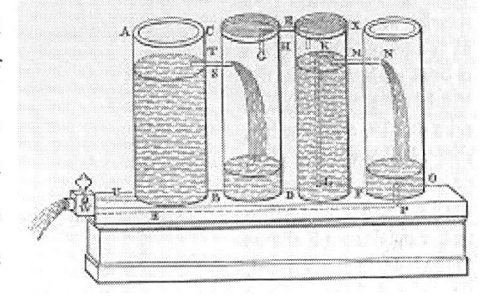

25. A Vessel from which Wine flows in proportion as Water is withdrawn

LET there be a vessel containing water, and a pipe in it provided with a key or cock, and let a figure float on the surface of the water; then if water, in any quantity, be drawn off through the pipe, wine shall flow from the figure in any given ratio to the water drawn off. Let A B (fig. 25) be the vessel of water, provided with a pipe, C, which admits of being closed; and on the surface of the water let a basin, D, float, in which is a perpendicular tube, E F, carved in the shape of some animal. Place near another vessel, G H, containing wine, in which is a bent siphon, K L M, one leg being within the vessel G H, and the other without, conducting into the tube E F. Now if we draw the wine through the lower mouth M, it will flow into the tube E F until the surface of the wine in the vessel G H and in the tube E F shall be at the same level. Let that level be in the line N X O P; and at the point P fix an open pipe, R. Hitherto there is no discharge of wine, but, if any quantity of water is drawn off through C, the basin D, and, with it, the tube E F will sink, and the surface of the wine [in the tube] will become lower than the surface N X; so that, the outer leg of the siphon being depressed, the wine will again pass on into the tube E F and run out through the pipe R. This will happen as often as we draw off water through the pipe C, the wine flowing in a fixed ratio to the water drawn off. The base of the vessel A B must bear the required proportion to the base of G H; and thus what was proposed is done.

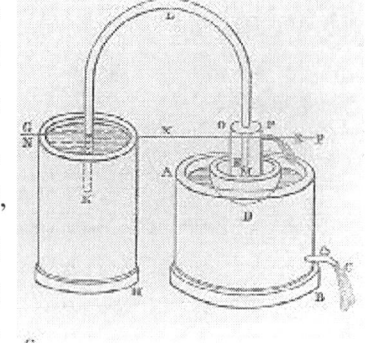

26. A Vessel from which Wine flows in proportion as Water is poured into another

IF it is required that the wine shall flow in a certain ratio to the water we pour into the vessel, we must proceed as follows. As before let A B (fig. 26) be the vessel containing water, and G H that which contains wine, but let the tube E F be outside the vessel A B. In A B let a ball, D, float, from which a cord, passing over a pulley, S, is attached to the tube E F so as to suspend it; and let all else correspond with what was stated in the last paragraph. The result will be that, when water is poured into the vessel A B, the ball D rising will lower the tube E F, and the wine will flow again. This may be effected in a different manner by attaching the cord from the ball D, across a third pulley, C, to another pulley, S, and across that again to the siphon K L. It will be found now that, when the ball rises, the siphon K L M, being suspended by the cord, is lowered, so that, the outer leg having again become the longer, the wine will flow through the mouth M.

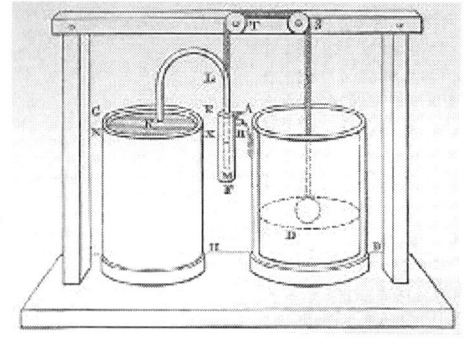

27. The Fire-Engine

THE siphons used in conflagrations are made as follows. Take two vessels of bronze, A B C D, E F G H, (fig. 27), having the inner surface bored in a lathe to fit a piston, (like the barrels of water-organs), K L, M N being the pistons fitted to the boxes. Let the cylinders communicate with each other by means of the tube X O D F, and be provided with valves, P, R, such as have been explained above, within the tube X O D F and opening outwards from the cylinders. In the bases of the cylinders pierce circular apertures, S, T, covered with polished hemispherical cups, V Q, W Y, through which insert spindles soldered to, or in some way connected with, the bases of the cylinders, and provided with shoulders at the extremities that the cups may not be forced off the spindles. To the centre of the pistons fasten the vertical rods S E, S E, and attach to these the beam A' A', working, at its centre, about the stationary pin D, and about the pins B, C, at the rods S E, S E. Let the vertical tube S' E' communicate with the tube X O D F, branching into two arms at S', and provided with small pipes through which to force up water, such as were explained above in the description of the machine for producing a water-jet by means of the compressed air. Now, if the cylinders, provided with these additions, be plunged into a vessel containing water, I J U Z, and the beam A' A' be made to work at its extremities A', A', which move alternately about the pin D, the pistons, as they descend, will drive out the water through the tube E' S' and the revolving mouth M'. For when the piston M N ascends it opens the aperture T, as the cup W Y rises, and shuts the valve R; but when it descends it shuts T and opens R, through which the water is driven and forced upwards. The action of the other piston, K L, is the same. Now the small pipe M', which waves backward and forward, ejects the water to the required height but not in the required direction, unless the whole machine be turned round; which on urgent occasions is a tedious and difficult process. In order, therefore, that the water may be ejected to the spot required, let the tube E' S' consist of two tubes, fitting closely together lengthwise, of which one must be attached to the tube X O D F, and the other to the part from which the arms branch off at S'; and thus, if the upper tube be turned round, by the inclination of the mouthpiece M' the stream of water can be forced to any spot we please. The upper joint of the double tube must be secured to the lower, to prevent its being forced from the machine by the violence of the water. This may be effected by holdfasts in the shape of the letter L, soldered to the upper tube, and sliding on a ring which encircles the lower.

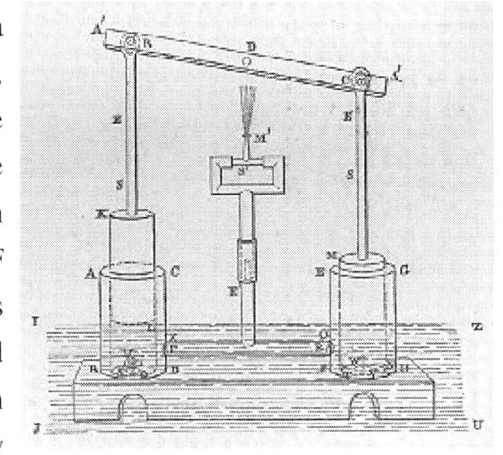

28. An Automaton which drinks at certain times only, on a Liquid being presented to it

IN any place provided with running water make a figure of some animal in bronze or any other material: when a cup is offered to it, the animal shall drink with a loud noise so as to present the appearance of thirst. The following is the construction. A B (fig. 28) is a vessel into which a stream of running water, C, falls. In A B place a bent siphon or inclosed diabetes, D E F, one leg of which must project below the bottom of the vessel. Underneath this let there be an air-tight pedestal, G H K L, also containing a bent siphon, M N X. Below the orifice F place a funnel, O P, the tube of which must descend into the pedestal leaving a passage for the water between its extremity and the bottom. Let the mouth of the animal be at R, from which a concealed tube, R S T, must run along one of the feet, or some other part, into the pedestal. When the vessel A B is filled, the water will overflow and run into the funnel, filling the pedestal G H K L and emptying the vessel A B; in like manner, when the pedestal is full, the water will overflow through the siphon M N X and empty the pedestal; and, as this becomes empty, the air will enter through the mouth R to fill up the void that is left. If, then, we apply a drinking vessel at R, the liquid will be violently attracted and sucked down instead of the air, until the pedestal within has become empty. Then the vessel A B is again filled and emptied, and the same will take place as before. In order that the Cup may be applied at the right time, that is, when the water is being drawn off from the pedestal, let something be contrived that will move when struck by water from the discharge through the siphon M N X. When this is seen to move, apply the drinking cup.

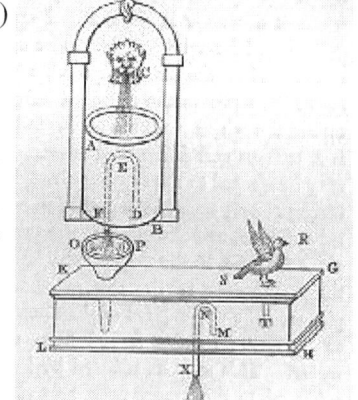

29. An Automaton which may be made to drink at any time, on a Liquid being presented to it

THERE is another way in which, by the aid of running water, the animal may be made to drink on the revolution of a carved figure of Pan. Let A B C D (fig. 29) be a pedestal, air-tight on every side, and divided into two chambers by a partition. On the surface place the animal, and let the tube E F G pass through its mouth. Within the pedestal, in the lower chamber, let there be a bent siphon, H K L, the lower leg projecting from the bottom: and let a funnel, M N, pass through the middle of the partition, its tube reaching nearly to the bottom. On the pedestal A B C D place another pedestal, O X, on which the figure of Pan, P R, is to stand, having attached to it the rod S which projects below into the pedestal. To S let the tube T U be fastened, at the end of which is the cup U Q, attached to and communicating with the tube. Let the tube be of such a length that, when the figure P R turns round, the cup U Q will be directly above the funnel M N. On the pedestal, and communicating with it, and directly above the funnel M N, place the cup W Y. Let the stream Z, (which must be greater than the discharge through the siphon H K L), flow into W Y: the liquid will pass through M N into the lower part of the pedestal, the contained air passing out through E F G: and now the pedestal will continue full as the influx is greater than the discharge. But, when we turn the figure P R round, the cup U Q will intercept the stream Z, which will pass elsewhere through the tube T U, and, as the water no longer flows into the lower chamber of the pedestal, the siphon H K L will empty it, and the air will enter through E F G. Thus, when the cup is applied, the animal will drink as before.

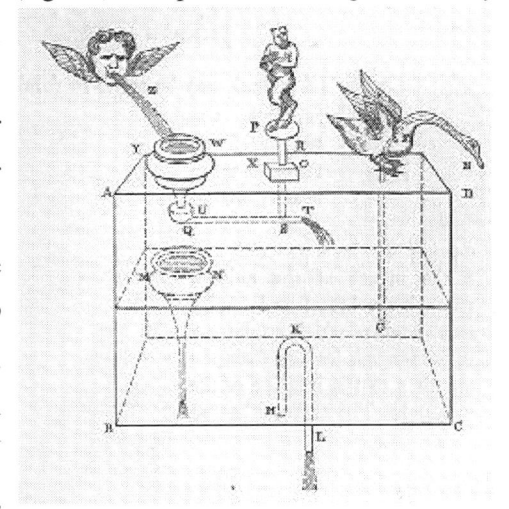

30. An Automaton which will drink any quantity that may be presented to it

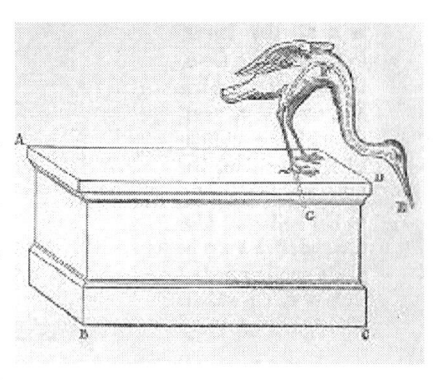

THE animal may be made to drink without the aid of running water, or of any thing to move the figure of Pan. Let A B C D (fig. 30) be a pedestal, and E the mouth of the animal, through the breast and hinder foot or tail of which a tube, E F G, is inserted, leading from the mouth E to the interior of the pedestal. The pedestal having been first firmly fixed, let a hole, E, so fine as to be scarcely discernible, be bored in the tube E F G which passes through the animal, in a line with the extremity G. Now if we fill the siphon E F G with water through some pipe above it, the mouth of which is applied to E, the siphon will continue full since its two orifices lie in the same level. If, therefore, a drinking vessel be brought to the mouth E, and a portion of the mouth immersed in it, it will be found that the leg of the siphon towards G has become the longer, so that it will attract the water, and the water attracted is carried into the pedestal A B C D. In this construction it is not necessary that A B C D should be air-tight.

31. A Wheel in a Temple, which, on being turned, liberates purifying Water

IN the porticoes of Egyptian temples revolving wheels of bronze are placed for those who enter to turn round, from an opinion that bronze purifies. There are also vessels of lustral (purificatory) water, from which the worshippers may sprinkle themselves. Let it then be required so to construct a wheel that, on turning it round, water shall flow from it to sprinkle the worshippers as we have described. Behind the entrance-pillar let a vessel of water, A B C D (fig 31), be concealed, having a hole, E, perforated in its base. Underneath the base let a small tube, F G H K be fastened, having also a hole bored opposite the orifice in the base and within this place another tube, L M, soldered to the tube F G H K at L, and opposite the orifice having in like manner a hole, S: between these two pipes let another pipe, N X O R, be closely fitted, with a hole at P opposite to E. Now, if the several holes are in one line, when water is poured into the vessel A B C D it will flow out through the pipe L M; but, if the pipe N X O R is made to revolve so as to change the position of the hole P, the discharge will cease. Attach the wheel to the pipe N X O R, and, if it is repeatedly made to revolve, water will flow out.

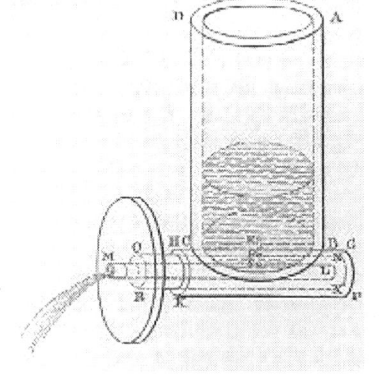

32. A Vessel containing different Wines, any one of which may be liberated by placing a certain Weight in a Cup

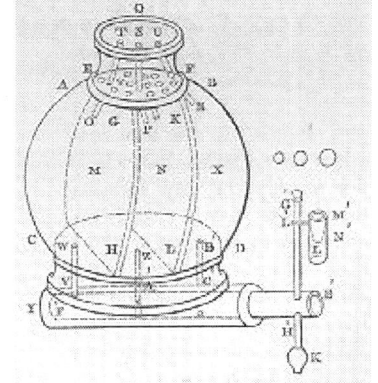

IF several kinds of wine be poured into a vessel by its mouth, any one of them may be drawn out through the same pipe: so that, if several persons have poured in the several wines, each one may receive his own according to the proportion poured in by him. Let A B C D (fig 32), be an air-tight vessel, the neck of which is closed by a partition, E F; and let the whole vessel be divided into as many compartments as we intend there shall be different kinds of wine. Suppose, for instance, that G H, K L, are the partitions, making three compartments, M, N, and X, into which the wine will be poured. In the partition E F pierce small holes, one in each compartment, O, P, R; and from these holes let small tubes, P S, O T, R U, communicating with the vessel, extend up into the neck. Perforate the partition E F, near each tube, with fine sieve-like holes, through which the liquid will pass into the compartments. When it is desired to pour in each kind of wine, place the fingers on S, T, and U, and pour in the wine through the neck Q; it will not pass into either of the compartments as the air contained in them has no outlet. But, if we set free one of the vents S, T, or U, the air contained in the corresponding compartment will pass out through the passage as the wine falls into the compartment. Then, placing the finger again on this vent, set another free in like manner, and pour in another kind of wine: and so in order with the rest, as many as there may be both of compartments and kinds of wine. We may procure each wine, in its due quantity, through the same pipe in the following manner. In the base of the vessel A B C D let there be tubes leading from each compartment, W Y from M, Z A' from N, and B' C' from X: the extremities of these tubes Y, A' and C', must communicate with another tube Y A' C', into which another tube, E' F', is tightly fitted, closed at the interior extremity F', and having holes pierced in it opposite to Y, A, and C' so that, as the tube E' F' revolves, when the holes pierced in it coincide consecutively with the holes Y, A', and C', they may admit the wine contained in each chamber and send it forth through the outer mouth of the tube E' F'. To the tube E' F' attach an iron rod, G' H'; to this, at the extremity U', solder a mass of lead, K', and at G' an iron pin, L' M', to the middle of which is fastened a cup, L, with the concavity upwards: let the interior of this be a hollow truncated cone of which M' is the larger circle and N' the less, and through this the pin L' M' is to pass. Take several balls of lead, varying in weight, and equal in number to the compartments M, N, X; and if we place the least of the balls in the cup M' N', it will descend by its weight until it touches the hollow surface of the truncated cone, causing the tube E' F' to revolve until the hole in it coincides with Y and admits the wine in the compartment M, which will flow as long as the ball remains in the cup, unless it be entirely exhausted: when we remove the ball the weight K' will turn back and close the orifice Y, and the discharge will cease. Again, insert another of the balls, and the cup will descend lower and turn the tube E' F' further round until the hole in it reaches the hole A', and then the wine in N will flow: as before when the ball is removed the weight K' will run down and close the orifice A', and the wine will cease to flow. If another ball still heavier be placed in the cup, the tube E' F' will be turned still further round, so that the wine in the compartment X will flow. It is necessary however that the least of the balls when placed in the cup should preponderate over the weight K', or, in other words, be able to cause E' F' to revolve; for then the other balls will preponderate and move E' F'.

33. A self-trimming Lamp

TO contrive a self-trimming lamp. Let A B C (fig. 33), be a lamp through the mouth of which is inserted an iron bar, D E, capable of sliding freely about the point E, and let the wick be wound loosely about the bar. Place near a toothed wheel F, moving freely about an axis, its teeth in contact with the iron bar, that, as the wheel revolves, the wick may be pushed on by means of the teeth. Let the opening for the oil be of considerable width, and when the oil is poured in let a small basin float upon it, G, to which is attached a perpendicular toothed bar, H, the teeth of which fit into the teeth of the wheel. It will be found that, as the oil is consumed, the basin sinks and causes the wheel F to revolve by means of the teeth of the bar, and thus the wick is pushed on.

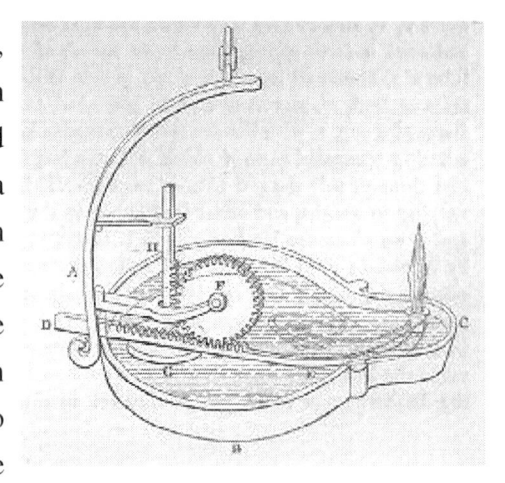

34. A Vessel from which Liquid may be made to flow, on any portion of Water being poured into it

IF into a vessel, provided at the bottom with an open spout, liquid is poured, the spout shall sometimes run from the first, sometimes when the vessel is half filled, and sometimes not until the whole is filled: in fine, when any proposed quantity of liquid has been poured in, the spout shall run until all is exhausted. Let A B (fig. 34), be the vessel, the neck of which is closed: insert the tube C D, air-tight, through the partition, and let it reach to the bottom of the vessel leaving only a passage for the water. Let E F G be a bent siphon the inner leg of which extends nearly to the bottom of the vessel while the other projects without, being fashioned in the shape of a water-spout: the curve of the siphon must be close to the neck of the vessel. In A B make an air-hole, H, near the partition and leading into the body of the vessel. If we intend the spout to run immediately on the entrance of the liquid, we must place the finger on the vent H, and the spout will run, for as the air in the vessel has no way of retreat, the liquid will rush out through the bent siphon. If we do not close H, the liquid will pass into the body of the vessel, and the spout cannot run until we again close the vent: and then, if we set the vent free, the siphon will exhaust all the liquid.

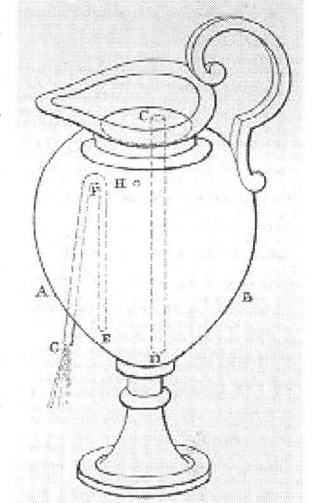

35. A Vessel which will hold a certain quantity of Liquid when the supply is continuous, will only receive a portion of such Liquid if the supply is intermittent

A VESSEL can be made which, as long as you pour in any liquid, admits it, but, if you once cease pouring, holds no more: the construction is in this manner. Let A B (fig. 35), be a vessel, the neck of which is closed by the partition C D. Through the partition insert the tube E F, reaching nearly to the bottom, and projecting above the partition so as almost to reach the brim of the vessel; and let this tube be encircled by another G H, the top of which is closed by a lid, at a sufficient interval from the partition and the tube E F to admit of the passage of water: in A B make an air-hole, K, leading into the body of the vessel. Now, if we pour liquid into the vessel's neck, it will be found that it will pass into the body through the tubes G H and E F, the air retreating through the vent K. But, if we cease pouring, and the neck of the vessel becomes empty, the air will break the continuity, so that any liquid in G H will flow down and fall upon the partition; for the breadth about the tube

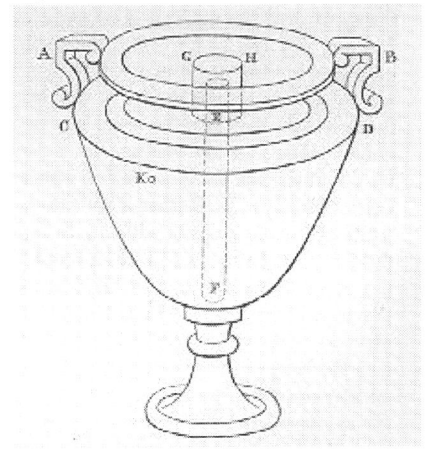

G H should be considerable, that the water may fall by its own weight. If more liquid be poured in, the air confined in the tubes E F and G H will not allow it to pass through, so that it will run over the brim of the vessel.

36. A Satyr pouring Water from a Wine-skin into a full Washing-Basin, without making the contents overflow

CONSTRUCT on a pedestal the figure of a satyr holding in his hands a wine-skin: place near a washing-basin, and into this let some liquid be poured until it is full; water shall be made to flow into the basin without running over, until all the water in the skin is exhausted. The following is the construction. Let A B (fig. 36), be a perfectly airtight pedestal, either cylindrical or octagonal in shape, as may seem more elegant, and divided into two chambers by the partition C D, through which the tube E F fitting closely into the partition, extends upwards nearly to the roof of the pedestal. Through the roof insert the tube G H, projecting slightly above the vessel, and lying exactly under the basin, while, below, it reaches to the bottom except that room must be left for the passage of water: this tube must be soldered into the roof of the pedestal and the partition. Another tube, K L M, must also be inserted through the roof, reaching not quite so low as the partition, soldered into the roof and carrying its stream into the basin, which lies above the tube G H and communicates with it. Now let the vessel A D be filled with water through an orifice N, which must be afterwards closed. If water is poured into the basin, it will pass through the tube G H into the vessel B C; and the air in B C, passing through the tube E F and into the vessel A D, will force the liquid in A D through K L M into the basin; and this being carried again into B C will force out the contained air as before, which, again, will force the water in the vessel A D into the basin: and this will go on until the water in A D is exhausted. The tube K L M must pass through the mouth of the skin and be particularly fine, that the display may last a considerable time.

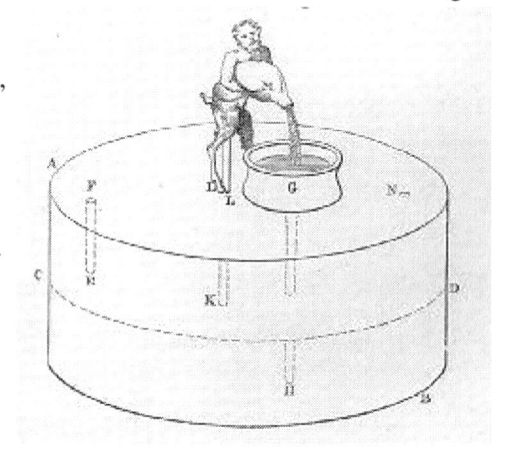

37. Temple Doors opened by Fire on an Altar

THE construction of a small temple such that, on lighting a fire, the doors shall open spontaneously, and shut again when the fire is extinguished. Let the proposed temple stand on a pedestal, A B C D (fig. 37), on which lies a small altar, E D. Through the altar insert a tube, F G, of which the mouth F is within the altar and the mouth G is contained in a globe, H, reaching nearly to its centre: the tube must be soldered into the globe, in which a bent siphon, K L M, is placed. Let the hinges of the doors be extended downwards and turn freely on pivots in the base A B C D; and from the hinges let two chains, running into one, be attached, by means of a pulley, to a hollow vessel, N X, which is suspended; while other chains, wound upon the hinges in an opposite direction to the former, and running into one, are attached, by means of a pulley, to a leaden weight, on the descent of which the doors will be shut. Let the outer leg of the siphon K L M lead into the suspended vessel; and through a hole, P, which must be carefully closed afterwards, pour water into the globe enough to fill one half of it. It will be found that, when the fire has grown hot, the air in the altar becoming heated expands into a larger space; and, passing through the tube F G into the globe, it will drive out the liquid contained there through the siphon K L M into the suspended vessel, which, descending with its weight, will tighten the chains and open the doors. Again, when the fire is extinguished, the rarefied air will escape through the pores in the side of the globe, and the bent siphon, (the extremity of which will be immersed in the water in the suspended vessel) will draw up the liquid in the vessel in order to fill up the void left by the particles removed. When the vessel is lightened the weight suspended will preponderate and shut the doors. Some in place of water use quicksilver, as it is heavier than water and is easily disunited by fire.

38. Other intermediate means of opening Temple Doors by Fire on an Altar

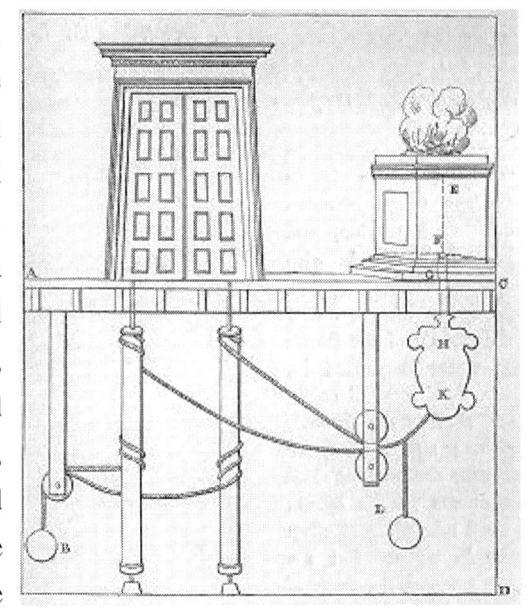

THERE is another way in which, on lighting a fire, the doors will open. As before, let a small temple stand upon a base, A B C D (fig. 38), on which is an altar, E. Let a tube, F G H, pass through the altar and be attached to a leathern bag, K, perfectly air-tight: beneath this let a small weight, L, hang, from which a chain is attached across a pulley to the chains round the hinges, so that, when the bag is folded together, the weight L preponderates and shuts the doors, and when fire is placed on the altar they are opened. For, as before, the air in the altar growing hot, and expanding, will pass through the tube F G H into the bag, and raise it up with the weight L; and then the doors will be opened. The doors will either open of themselves, as the doors of baths shut spontaneously, or they may have a counterbalancing weight to open them. When the sacrifice is extinguished, and the air which has entered the bag passes out, the weight, descending with the bag, will tighten the chains and close the doors.

39. Wine flowing from a Vessel may be arrested on the Introduction of Water, but, when the Supply of Water ceases, the Wine flows again

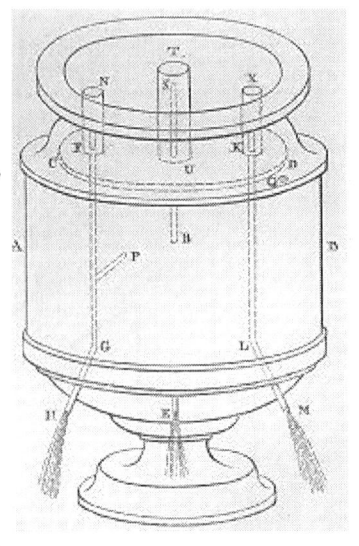

IF there be a vessel containing wine, and provided with three spouts, wine shall flow through the middle of the three; and, when water is poured in, the stream of wine shall cease, and water shall flow through the other two; again, when the stream of water ceases, wine shall flow through the middle spout: and this shall take place as often as we pour in water. Let A B (fig. 39), be a vessel the neck of which is closed by the partition C D, and having a spout, E, at the bottom. Let two tubes, F G H, K L M, terminating in spouts, pass through the partition and project above it; and round the projecting parts place other tubes, N, X, covered with lids at the top and extending to the partition except a passage for the water. Another tube, P, reaching nearly up to the partition, communicates with F G H. Having first closed the spout E, fill the vessel A B with wine through an orifice, Q, which must be carefully closed afterwards. When E is set free it will be found that wine flows through it, for air enters from without into the void created, through the orifice H and the tube P.

Now, if we pour water upon the partition C D, it will be carried out through the tubes F G H, K L M; but, as the air has no means of entering the vessel, A B, the wine will cease to flow until all the water has escaped, when the air finds an entrance again and the wine flows. Instead of the tube P, another tube, R S, may be used, piercing through the partition, about which another, T U, must lie, like the tubes N and X, but higher than those, so that R S may rise above the lip of the vessel. The same result will follow.

40. On an Apple being lifted, Hercules shoots a Dragon which then hisses

ON a pedestal is placed a small tree round which a serpent or dragon is coiled; a figure of Hercules stands near shooting from a bow, and an apple lies upon the pedestal: if any one raises, with the hand, the apple a little from the pedestal, the Hercules shall discharge his arrow at the serpent and the serpent hiss. Let A B (fig. 40) be the proposed pedestal, air-tight and divided by a partition, C D. Fixed in the partition is a hollow truncated cone, E F, the lesser circle of which, F, is open and approaches to the bottom of the pedestal, leaving a sufficient interval for the passage of water. To this cone must be tightly fitted another cone H, attached by means of a chain through a hole in the surface, to the apple K, which lies on the pedestal. Let the Hercules hold a small bow of horn, the string of which is stretched, and at the proper distance from the hand. In the right hand, and directed towards the serpent, let there be a hand in every respect similar to the visible hand, but smaller, and holding the trigger. From the extremity of the trigger let a chain, or cord, proceed through the pedestal and be attached to a pulley, which is placed above the partition, and again to the chain which is connected with the cone and apple. Now we must draw the bow, and placing the trigger beneath the hand, close it so that the cord is stretched and draws the apple tightly downwards: the cord must run inside the Hercules and through the body and hand. From the partition let a small tube, one of those which are used to whistle, extend above the pedestal and pass under the tree or along its trunk. Then fill the vessel A D with water. Let L M be the tree, N X the bow, S P the string, R S the hand that grasps the bow, T U the trigger, Q W the cord, W the pulley round which the cord runs, and Y Z the whistling pipe. Now if some one raise the apple K, he will at the same time raise the cone H, tighten the cord Q W, and draw back the hand, so that the arrow is discharged: and the water in A D, being carried into B C, will drive out the air contained in B C through the pipe, and produce the hissing sound. When the apple is replaced, the cone H fitting again into the other, will stop the stream of water so that no sound is produced.

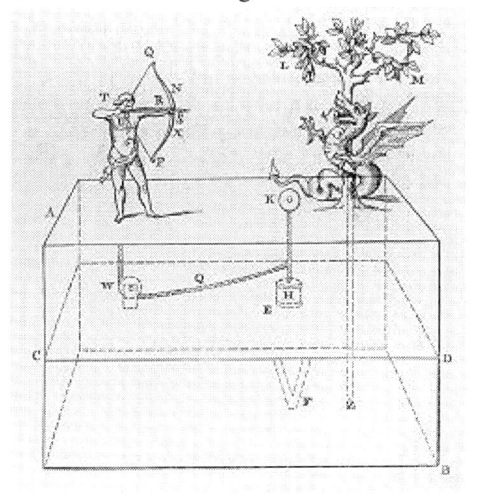

We must now re-arrange the arrow and leave it. If the vessel B C is full, it can be emptied again by means of a spout with a key: A D must be filled as before.

41. A Vessel from which uniform Quantities only of Liquid can be poured

THE following is the construction of the vessel called a dicaeometer, which, having been filled with liquid, discharges an equal quantity every time it is inverted. Let A B (fig. 41), be a vessel the neck of which is closed by the partition A B: near its bottom let there be a small globe, C, holding the measure of water we intend to flow out. Through the partition insert a small and very fine tube, D E, communicating with the globe. In the lower part of the globe perforate a small hole, F, from which a pipe, F G, extends upwards, running just beneath, and communicating with, the handle of the vessel which is hollow. Near the hole just mentioned make another at L towards the body of the vessel: the handle also must have a vent at H. Having first stopped the vent H we must fill the vessel with liquid through a hole which must afterwards be carefully closed, or the vessel may even be filled through the tube D E. Now, if we invert the vessel and set the vent H free, the liquid in the globe C and the tube D E will flow out. If we again close the vent and restore the vessel to its original position, the globe and tube will be filled again, for the air they contain will be driven out by the liquid rushing in; and, when the vessel is once more inverted, a like quantity of liquid will again flow out, except indeed with some difference as to the tube D E, for it will not be always filled, but as the vessel grows empty it will be empty itself: this difference however is extremely small.

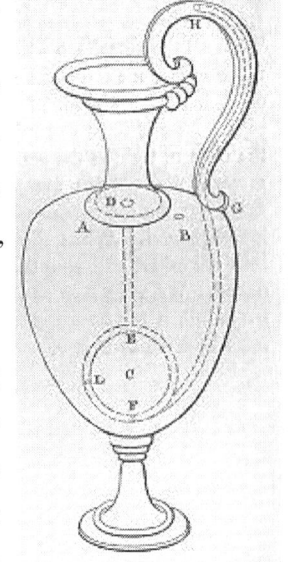

42. A Water-Jet actuated by compressed Air from the Lungs

HERE are vessels from which water is forced up by blowing into them. Through the neck of the vessel (fig.42), a tube is inserted, reaching nearly to the bottom, and soldered in at its mouth. Stop this mouth with the finger, and pour in some liquid through a hole: then, having blown into the vessel through the same hole, close it by means of a key, and set free the mouth of the tube; the liquid will be made to spout up through the orifice by the compressed air which was blown in.

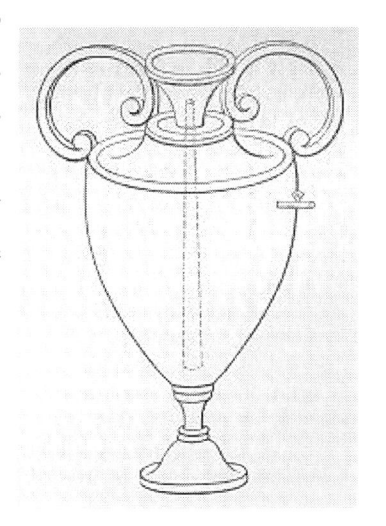

43. Notes from a Bird produced at intervals by an intermittent Stream of Water

THE notes of birds are produced at intervals as follows. Take an air-tight vessel (fig. 43), through which a funnel is inserted, the tube being far enough from the bottom of the vessel to allow of the passage of water. Above the funnel is placed a hollow vessel, turning on pivots, and having a weight below, into which water is continually carried. So long as the vessel on the pivots is empty it will be found to remain upright, for a weight is attached to its bottom; but, when the vessel is filled the water is overturned into the air-tight vessel, and the air contained in the vessel being driven out through a small pipe will produce the sound. The vessel is emptied of water by means of a bent siphon, and, while it is being emptied, the vessel on pivots is again filled and overturned. It will be requisite that the stream of water should not fall into the centre of the vessel on pivots, that when filled it may be inverted speedily.

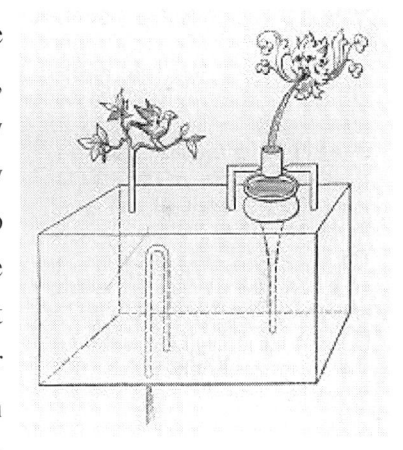

44. Notes produced from several Birds in succession, by a Stream of Water

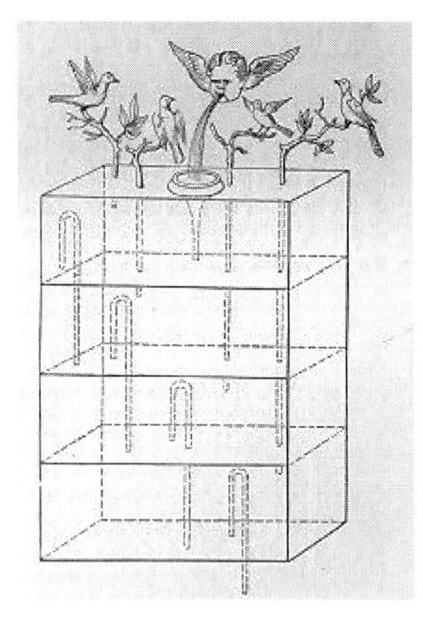

SOUNDS are produced at intervals in another way as follows. A vessel is taken (fig. 44), provided with several transverse partitions. In the chambers are placed siphons conducting into the chambers beneath, the streams through them being unequal. In the lower compartment is placed the pipe which produces the sound, and the stream of water falls into the upper compartment. It will be found that when the upper chamber is filled, the water passes through the siphon placed there into the chamber below, until it has arrived at the lowest, and the vessel being air-tight, the air in this chamber is driven out through the pipe and produces the sound.

45. A Jet of Steam supporting a Sphere

BALLS are supported aloft in the following manner. Underneath a cauldron (fig. 45), containing water and closed at the top, a fire is lighted. From the covering a tube runs upwards, at the extremity of which, and communicating with it, is a hollow hemisphere. If we put a light ball into the hemisphere, it will be found that the steam from the cauldron, rising through the tube, lifts the ball so that it is suspended.

46. The World represented in the Centre of the Universe

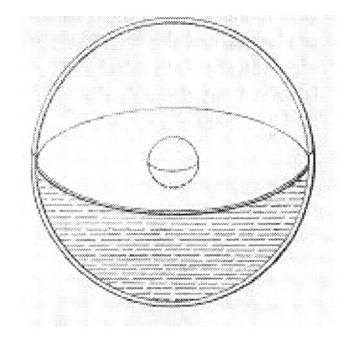

THE construction of a transparent globe containing air and liquid, and also of a smaller globe, in the centre, in imitation of the world. Two hemispheres of glass are made (fig. 46): one of them is covered with a plate of bronze, in the middle of which is a round hole. To fit this hole a light ball, of small size, is constructed, and thrown into the water contained in the other hemisphere: the covered hemisphere is next applied to this, and, a certain quantity of liquid having been removed from the water, the intermediate space will contain the ball; thus by the application of the second hemisphere what was proposed is accomplished.

47. A Fountain which trickles by the Action of the Sun's Rays

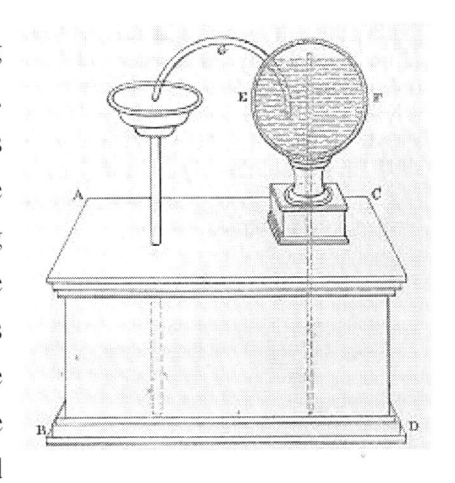

THE "fountain" as it is called may be made to trickle as long as the sun falls upon it. Let there be an airtight pedestal, A B C D (Fig. 47), through which a funnel is inserted, its tube extending within a very little of the bottom. Let E F be a globe, from which a tube leads into the pedestal, (reaching nearly to the bottom of the pedestal and to the circumference of the globe,) while a bent siphon, fitted into the globe, leads into the funnel. Now pour water into the globe; and when the sun falls upon the globe, the air in it, being heated, will drive out the liquid, which will be carried along the siphon G, and pass through the funnel into the pedestal. But when the globe is in the shade, the air having escaped through the globe, the tube will again suck up the liquid, and fill the void which had been produced; and this will take place as often as the sun falls upon the globe.

48. A Thyrsus made to whistle by being submerged in Water

BY immersing a thyrsus in water to produce the sound either of a pipe or of any bird. Let A B C D (fig. 48), be a thyrsus; and at the extremity of its head, which must be hollow and shaped like a fir-cone, let there be an orifice D. Close the shaft a little below the mouth by the partition A E, and place near it a small pipe, F, just beneath the mouth of the tube, and passing through an orifice in the partition. If we insert the thyrsus in water and force it downwards, the air contained in it being driven out by the water will produce a sound. If there is nothing but the pipe we shall have a whistle only; but if there is any quantity of water under the partition there will be a gurgling sound.

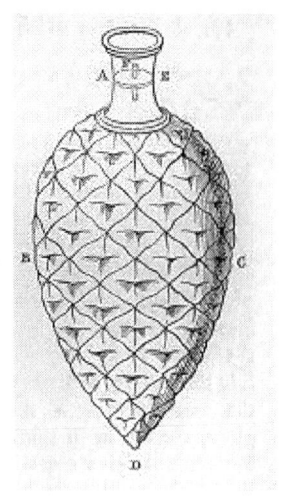

49. A Trumpet, in the hands of an Automaton, sounded by compressed Air

A FIGURE stands upon a pedestal having a trumpet in its mouth: if it be blown into, the trumpet shall sound. Let A B C D (fig. 49), be an air-tight pedestal on which a figure stands, and within the pedestal let there be a hollow hemisphere, E F G, covered over at the top and having small holes in the bottom. From the hemisphere a tube, H F, extends upwards into the figure in the direction of the trumpet, which is provided with a mouth-piece. Pour liquid into the pedestal through a hole which must be afterwards stopped again by means of a valve or tap called a smerisma. Now, if we blow into the bell of the trumpet, the air passing from us will force out through the holes the water in the hemisphere, which will mount up into the pedestal: but when we withdraw the breath, the water will enter the hemisphere again and force out the air, which, passing out through the mouthpiece, will produce the sound of a trumpet.

50. The Steam-Engine

PLACE a cauldron over a fire: a ball shall revolve on a pivot. A fire is lighted under a cauldron, A B, (fig. 50), containing water, and covered at the mouth by the lid C D; with this the bent tube E F G communicates, the extremity of the tube being fitted into a hollow ball, H K. Opposite to the extremity G place a pivot, L M, resting on the lid C D; and let the ball contain two bent pipes, communicating with it at the opposite extremities of a diameter, and bent in opposite directions, the bends being at right angles and across the lines F G, L M. As the cauldron gets hot it will be found that the steam, entering the ball through E F G, passes out through the bent tubes towards the lid, and causes the ball to revolve, as in the case of the dancing figures.

51. A Vessel from which flowing Water may be stopped at pleasure

IF a bowl stands upon a pedestal and has an open waterspout, the discharge shall suddenly cease, though there be no slide or tap attached to shut the spout. Let A B (fig. 51), be the bowl on the pedestal C: through the bottom of the bowl and the pedestal insert a tube, D E F, terminating in a spout; and at the handle of the vessel fix a bar, G H, against which another bar, K L, may move about the pin H: at the extremity K place a vertical bar, K M, moving about the pin K: to this bar let a box, N X, be attached at M, having weight, and large enough to inclose the tube D E F. When the bowl is full, if we depress the extremity L of the bar, the box N X will ascend, and, when this is raised, the water in the bowl will be carried out through the tube D E F: but if the extremity L be set free, the box will descend and encompass the tube D E F, and the air it contains, having no way of escape, will disconnect the liquid round the tube D E F, and prevent it from being further carried out through the mouth D. When we again depress the extremity L the spout will run as before.

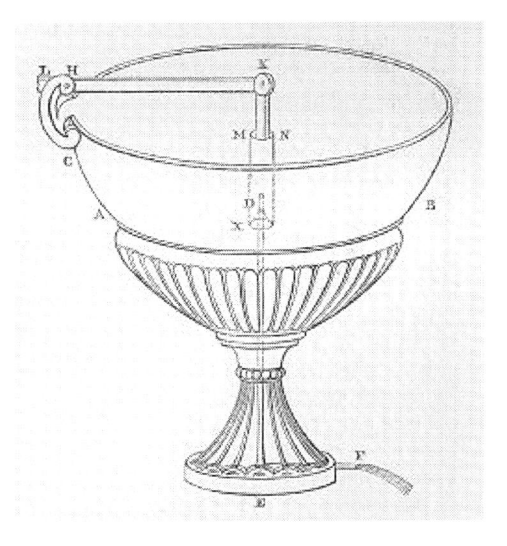

52. A Drinking-Horn in which a peculiarly formed Siphon is fixed

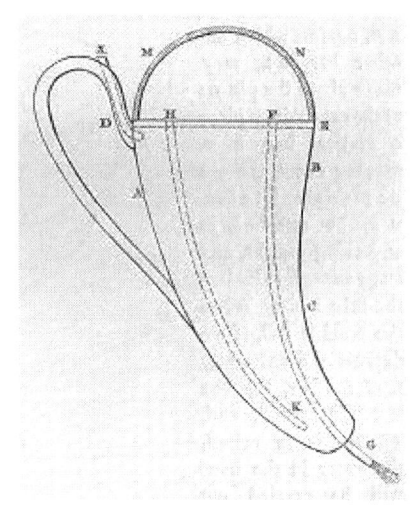

THE construction of a drinking-horn such that, if a cover of glass be placed upon it, while a discharge is going on from the vessel, the liquid shall ascend into the glass cover and be thrown back. A B C (fig. 52), is a drinking horn, closed by the covering D E; and from D E extend two tubes, F G, H K, one of them, H K, leading into the interior of the vessel, the other, F G, leading outside. A glass cover, M N, incloses this; and in the top, D E, outside the glass vessel, is an aperture, X, through which water may be poured. When the horn is filled through this aperture, the tube H K will be filled at the same time, and as the water is poured in it will ascend into the glass vessel so as to be carried outside through the tube F G. Thus we shall have the arrangement of a bent siphon, of which H K is the smaller leg and F G the greater, so that it will attract the liquid in the horn as it ascends into the cover; it will also attract the air contained in the cover, which is lighter than the liquid, and the water will appear to be thrown back into the void space left by the air and to descend by its own weight; for this upward motion is contrary to its nature.

53. A Vessel in which Water and Air ascend and descend alternately

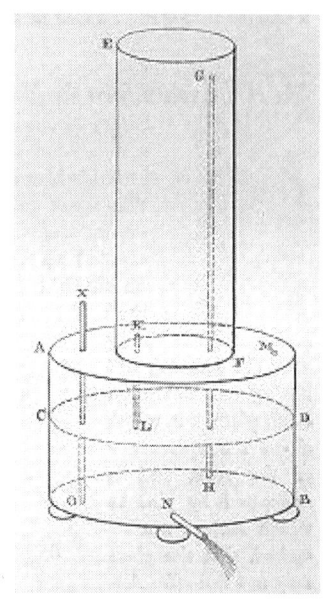

THERE is also another contrivance by which liquid is borne steadily upwards and remains, so as to seem perpetually ascending. Let A B (fig. 53), be a perfectly air-tight pedestal, partition, C D, and a cover, E F, also perfectly air-tight. In the cover E F let there be a tube, G H, reaching nearly to the top, and passing through an orifice in the partition C D, and another tube, K L, passing through the top of the pedestal but not descending quite so low as the partition. In the pedestal, and outside the glass cover, let there be an aperture, M, through which the vessel A D is to be filled, and near the bottom of the pedestal a spout, N; also one other tube, X O, passing through the partition and reaching nearly to the bottom of the pedestal, through which the vessel C B may be filled. If the spout, N, be closed the air in C B will pass out through the tubes G H, K L, and the hole M; and when C B is full we must fill A D through the hole M, for the air contained in it will pass out through the same hole. Now, if we set the spout N free, the air in the glass cover will pass through the tube G H into the void space left in C B, and water will ascend from A D through the tube K L into the void space left in the cover, while into the void of the vessel A D air will enter through the aperture M and this will go on until the glass cover is filled: but the spaces A D, C B, E F, must be of equal capacity that the air and water may take the place of one another. When C B is exhausted and the continuity of the air is broken, the water will again descend out of the glass cover into A D, air passing into the cover through the spout N and the tube G H. The air in A D will pass out through the aperture M.

54. Water driven from the Mouth of a Wine-skin in the Hands of a Satyr, by means of compressed Air

IF wind is blown through the mouth of certain figures, they spout up water through some other place. For example, if a satyr holds a wine skin, water shall be spouted up through the skin. A B C D (fig. 54), is an air-tight pedestal on which the figure is placed; through the mouth of the figure a tube, E F is inserted, communicating with the pedestal, and having underneath it a small plate, G H, which closes the aperture F of the tube, and is supported by pins to which buttons are attached, that the plate may not fall off. Another tube, K L, is passed through the pedestal, of which the extremity, K, must be contiguous to the point at which the water-jet is to be, and the extremity, L, reach to the bottom of the pedestal, leaving only a passage for the water. At the extremity K there must be a valve or tap by which the aperture K, which is very small, may be shut. Now if we pour any quantity of water into the pedestal through a hole, which we must afterwards stop, and, having closed the aperture K, blow in air through the tube E F, the air blown in will thrust aside the plate and descend into the pedestal: and, if this is done several times, the air in the pedestal will be compressed and close the plate. Let the valve or tap be opened, and after a short time the compressed air will drive the liquid in the pedestal violently out through the aperture K, until all the liquid is spouted up, and the air is brought back to its natural state, that is, in which it is no longer subject to compression.

55. A Vessel, out of which Water flows as it is poured in, but if the supply is withheld, Water will not flow again, until the Vessel is half filled; and on the supply being again stopped, it will not then flow until the Vessel is filled.

THERE are some vessels which, when water is poured in, flow immediately, but, if we discontinue pouring for a short space, do not flow again, though water is poured in afresh, till they are half full, when they begin to flow once more; and if we discontinue again, do not flow any more till they are quite full. Let A B (fig. 55), be a vessel containing, concealed in its interior, three siphons, C, D, E, one leg of each being near the bottom of the vessel, while the other, fashioned into a water-spout, conducts outside the vessel. At the outer extremities of the siphons, apply vessels, F, G, H, the bottoms of which are far enough from the orifices of the siphons to admit the passage of water between; and let all this be encompassed by another vessel, as it were a pedestal, K L M N, which is provided with a spout at X. Let the bend of the siphon C be close to the bottom of the vessel A B; that of D, half way up its height, and that of E, near the neck. Now, if we pour water into the vessel A B, it will immediately flow through the siphon C since its bend is near the bottom: but, if we cease pouring, the liquid poured in will be drawn off through the pipe F, and the vessel F will be found full of water, while the other part of the siphon C will be full of air. Consequently, when liquid is again poured into the vessel, it will not pass through the siphon C, owing to the air which is contained in the siphon between the water which is being poured in and that in the vessel F. The liquid will therefore rise as high as the bend of the siphon D, which is at the middle of the vessel, and then it will begin to flow: but, if we again cease pouring, the same will happen as has been explained in the case of the siphon C. A like result must be imagined with the siphon E. It will be necessary to pour in the stream gently, that the air intercepted in the siphon may not be forcibly driven out.

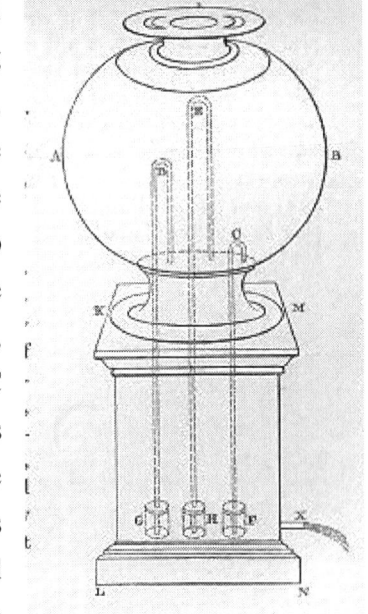

56. A Cupping-Glass, to which is attached, an Air-exhausted Compartment

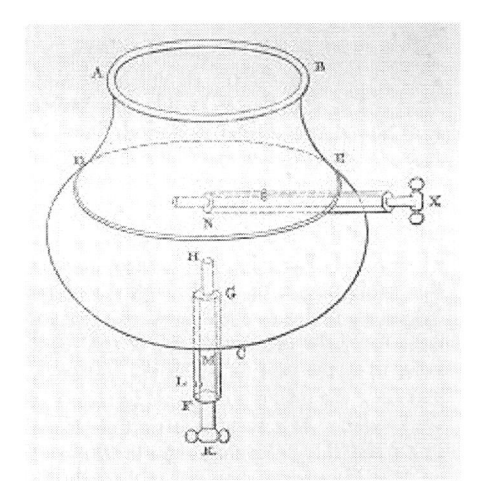

THE construction of a cupping-glass which shall attract without the aid of fire. Let A B C (fig. 56), be a cupping-glass, such as is usually applied to the body, having a partition across it, D E: through the bottom of the cupping-glass let two sliding tubes be inserted, F G being the outer tube and H K the inner; and in these, but outside the cupping-glass, pierce corresponding holes, L and M. Let the inner extremities of both the tubes be open, but the outer extremity of H K be closed and provided with a handle. Under the partition D E place another pair of sliding tubes, N X, like those just described; but the corresponding holes must be within the cupping-glass, and be precisely adapted to a hole in the partition. When these perforations are complete, let the handles of the sliding tubes be turned round, so that the holes in the lower tubes may be in a line, while those under the partition, not being allowed to coincide, remain closed. Now, the chamber D C being full of air, by applying the orifice L M to the mouth we can suck out a portion of that air; and then, by turning the handle again and not removing the tubes from the mouth, we can keep the air in the vessel C D rarefied; and this must be repeated until we have drawn off a large quantity of air. Then, applying the glass to the flesh in the usual manner, we open the holes in the sliding tubes N X by means of the handle; and it must follow that some of the air in the vessel A D E will pass into the place of the air withdrawn from C D, while into the void thus created both the flesh and the matter about it will be drawn up through the interstices of the flesh which we call invisible spaces or pores.

57. Description of a Syringe

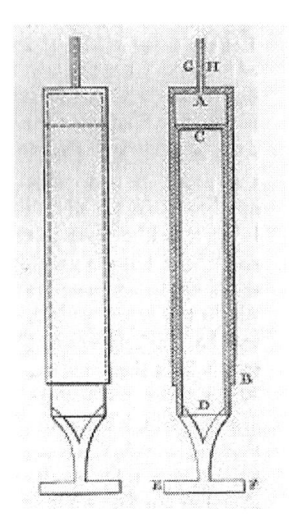

THE instrument called a pyulcus acts on the same principle. A hollow tube, of some length, is made, A B (fig. 57); into this another tube, CD, is nicely fitted, to the extremity C of which is fastened a small plate or piston, and at D is a handle, E F. Cover the orifice A of the tube A B with a plate in which an extremely fine tube, G H, is fixed, its bore communicating with A B through the plate. When we desire to draw forth any pus we must apply the extreme orifice of the small tube, H, to the part in which the matter is, and draw the tube C D outwards by means of the handle. As a vacuum is thus produced in A B something else must enter to fill it, and as there is no other passage but through the mouth of the small tube, we shall of necessity draw up through this any fluid that may be near. Again, when we wish to inject any liquid, we place it in the tube A B, and, taking hold of E F, depress the tube C D, and force down the liquid until we think the injection is effected.

58. A Vessel from which a Flow of Wine can be stopped, by pouring into it a small Measure of Water

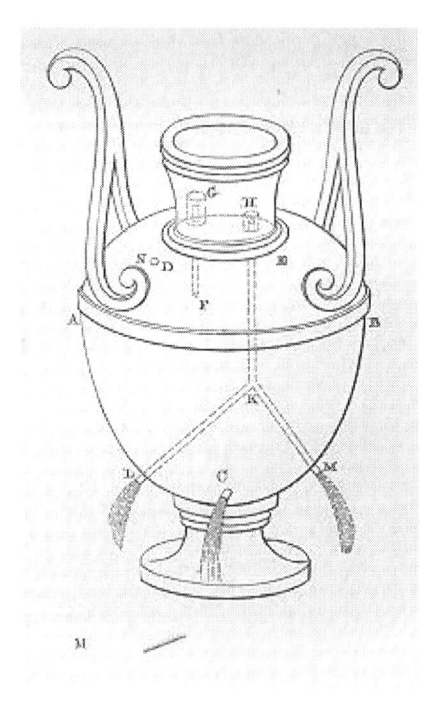

IF there be a vessel full of wine and provided with a running spout, when a cyathus, or small measure, of water is poured upon the neck of the vessel, the discharge of wine shall cease, but, if a second measure of water poured on, this last shall flow out with the former, or the two measures of water shall flow out through two different spouts; and, after all the water is drawn off, the wine shall flow again from the centre spout: moreover, this shall happen as often as any liquid is poured on and flows out. Let A B (fig. 58), be a vessel with a spout, C, at the bottom, and closed at the neck by the partition D E from which extends a tube, F G, encircled by another tube which is sufficiently removed from the partition to allow of the passage of water, as in the case of the inclosed diabetes. Through the partition insert another tube, H K, projecting to a less height above the partition than the former tube, and branching off below into two spouts L and M; and let this tube also be encircled by another tube distant a small space from the partition: furthermore let the vessel have a vent N just under the partition. Now, if, after closing the spouts, we pour in the wine, it will pass into the body of the vessel through the tube F G, for the air will escape through the vent N: but when we close the vent and set the spouts free, the liquid intercepted in the tube H K will flow through L and M, and that contained in the vessel through C. If, however, while C is still running, we pour a small measure of water upon the partition, the air will no longer be able to enter through F G, and the discharge through C will cease: but if a second measure is poured on, the water will rise above the tube H K, and be carried through into the spouts L and M, the whole being drawn up; and then, the tube F G being opened to the air will enable the spout C to flow as before. This result will take place as often as we pour on the measures of water.

59. A Vessel from which Wine or Water may be made to flow, separately or mixed

FROM a vessel full of pure wine sometimes the wine flows; if water is poured in, pure water flows out; then again pure wine; and, if it is desired, when the water is poured in a mixture shall be discharged. Let A B (fig. 59), be a vessel, having a partition near the neck, C D, through which a tube, E F, is inserted, passing out below and terminating in a spout. In the tube E F, within the vessel and near the bottom at G, let there be a fine hole, and a vent under the neck at H. Now, if we close the spout F, and pour in the wine, it will pass into the body of the vessel, the air escaping through the vent H but if we stop the vent and set the spout free, nothing will flow out except what is intercepted in the tube E F. If water is then poured in, it will flow out pure, and, when the vent is set free, a mixture is discharged: if nothing more is poured in, pure wine will flow.

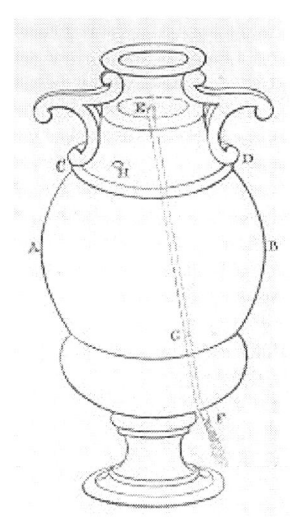

60. Libations poured on an Altar, and a Serpent made to hiss, by the Action of Fire

WHEN a fire is kindled on an altar, figures placed near shall offer libations, and a serpent hiss. Let there be a hollow pedestal, A B (fig. 60), on which is an altar, C, containing within it a tube, D E, which descends from the hearth of the altar to the pedestal, and then branches off into three tubes, E F leading to the mouth of the serpent; E G H to a wine vessel K L, (the bottom of which must be higher than the figure M,) and fastened to the lid of K L cross-bar fashion; while the other tube E N X, in like manner, extends into another wine vessel O P, also terminating in a cross-head. Both these tubes must be soldered into the bottoms of the vessels, and in each wine vessel there must be a bent siphon, R S, and T U, one extremity of each being immersed in the wine, and the other, (from which extend the hands of the figures which are to pour the libations,) passing, air-tight, through the sides of the wine vessels. When the fire is about to be kindled, pour first a little water into the tubes, that they may not be burst by the dry heat, and close up everything that no air may pass through. The hot air, becoming mixed with the water, will ascend along the tubes to the cross-heads, and through them it will exert pressure on the wine, and carry it to the bent siphons R S and T U. The wine flowing through the hands of the figures produces a libation as long as a fire is burning on the altar. The other tube, conveying the hot air to the mouth of the serpent, will cause the serpent to hiss.

61. Water flowing from a Siphon ceases on surrounding the End of its longer Side with Water

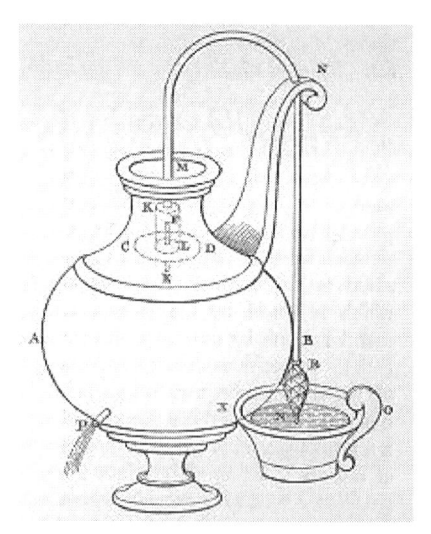

LET there be an air-tight vessel provided with an open spout, and by its side a thyrsus under which is a cup full of water: if the cup is removed, as long as it is withdrawn, a small stream shall flow from the mouth; but when the cup is pushed back, the spout shall run no longer. Let A B (fig. 61), be the vessel described, having its neck closed by the partition CD; from CD, and fitted air-tight in it, a tube, E F, extends, about which lies another tube, K L, forming an inclosed diabetes. With K L another tube, M N, communicates, of which the mouth M is open, while the outer leg is placed in a cup, O X, into which water has been poured until it is full; it is clear that so much of the leg of the siphon as is in the cup will be filled at the same time. Into the neck of the vessel A

B a little water must be poured, just enough to close all entrance for the air; and, when A B is full, the spout P, though open, will not run, since the air has no means of entrance, because of the water poured into the neck. But if the cup is drawn slightly downwards, some portion of the leg of the siphon which is in the cup must be emptied, and into the part emptied the contiguous air will be drawn: this air will attract some of the water which was poured into the neck, so that the water shall rise above the mouth F; and hence, the air having found an entrance, the spout P will run until the cup O X is pushed up again, causing the water to return to its old position and to close the passage for the air so that the spout will cease to flow. This will happen as often as the cup is withdrawn and applied: it is necessary however that the cup be not wholly drawn away, that the siphon leg may not be wholly emptied. Let the tube M N be fashioned like a thyrsus, R N being its shaft: thus the spectacle will be properly arranged.

62. A Vessel which emits a Sound when a Liquid is poured from it

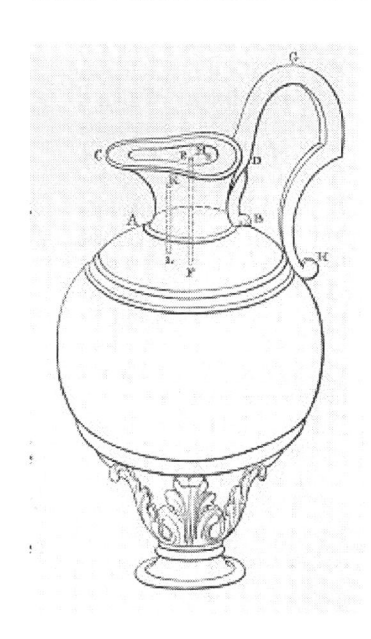

THE construction of a flagon which utters a sound when liquid is forced from it. Take a flagon (fig. 62), such as is about to be described, the neck of which is closed by the plate A B, and the mouth by C D; and through both these partitions, fitting into them air-tight, let a tube, E F, be inserted. G H is the handle of the flagon, and K L a tube placed in the opposite side of the neck, fitting closely into the partition A B and far enough distant from C D to allow of the passage of water: in C D let there be a small pipe M such as will utter sound. The flagon may be filled through the tube E F, the air passing out through the tube K L and the pipe M; and if we take the handle of the flagon and incline it so as to pour out the contents, water will flow out of the vessel through the tube E F, and into the neck B C through K L the air contained in the neck being forced out through M gives forth a sound. There should be another hole in A B through which air may pass again when the vessel is righted.

63. A Water-Clock, made to govern the quantities of Liquid flowing from a Vessel

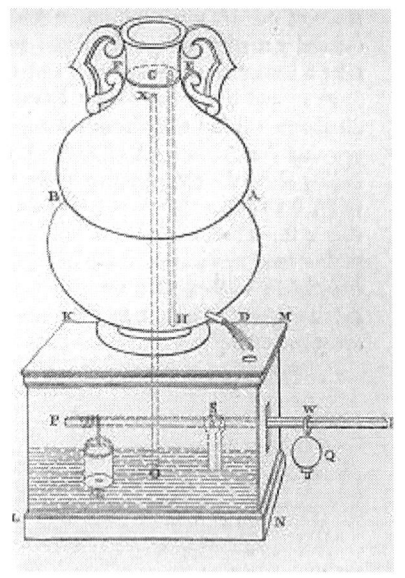

A VESSEL containing wine, and provided with an open spout, stands upon a pedestal: it is required by shifting a weight to cause the spout to pour forth a given quantity, - sometimes, for instance, a half cotyle (1/4 pint), sometimes a cotyle (1/2 pint), and, in short, whatever quantity we please. A B (fig. 63), is the vessel into which wine is to be poured: near the bottom is a spout, D: the neck is closed by the partition E F, and through E F is inserted a tube, G H, reaching nearly to the bottom of the vessel, but so as to allow of the passage of water. K L M N is the pedestal on which the vessel stands, and O X another tube reaching within a little of the partition and extending into the pedestal, in which water is placed so as to cover the orifice O, of the tube. Fix a rod, P R, one half within, and the other without, the pedestal, moving like the beam of a lever about the point S; and from the extremity P of the rod suspend a water-clock, T, having a hole in the bottom. The spout D having been first closed, the vessel should be filled through the tube G H before water is poured into the pedestal, that the air may escape through the tube X O: then pour water into the pedestal, through a hole, until the orifice O is closed, and set the spout D free. It is evident that the wine will not flow, as there is no opening through which air can be introduced: but if we depress the extremity R of the rod, a portion of the water-clock will be raised from the water, and, the vent O being uncovered, the spout D will run until the water suspended in the water-clock has flowed back and closed the vent O. If, when the water-clock is filled again, we depress the extremity R still further, the liquid suspended in the water-clock will take a longer time to flow out, and there will be a longer discharge from D: and if the water-clock be entirely raised above the water, the discharge will last considerably longer. To avoid the necessity of depressing the extremity R of the rod with the hand, take a weight Q, sliding along the outer portion of the rod, R W, and able, if placed at R, to lift the whole water-clock; if at a distance from R, some smaller portion of it. Then, having obtained by trial the quantities which we wish to flow from D, we must make notches in the rod R W and register the quantities; so that, when we wish a given quantity to flow out, we have only to bring the weight to the corresponding notch and leave the discharge to take place.

64. A Drinking-Horn from which a Mixture of Wine and Water, or pure Water may be made to flow alternately or together, at pleasure

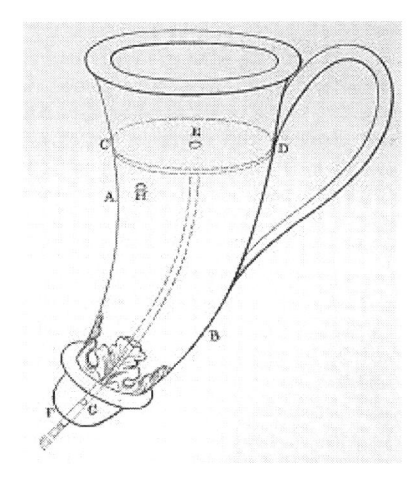

THE construction of a drinking-horn from which at first a mixture shall flow; when we please, on pouring in water, water alone, and then again a mixture. Let A B (fig. 64), be a drinking-horn, its neck closed by the plate C D, through which is inserted a tube, E F, leading to the orifice F, and having a hole, G, bored in it within the vessel: in the vessel just under the partition make a vent H. Now, if we close the orifice F and pour in the mixture, it will pass into the body of the vessel through the hole G; and if we set F free, the mixture will flow through it, the air escaping by the vent H. Again, if we close H and pour in pure water, the mixture will no longer flow as the air has no means of entrance, but pure water; and, when H is set free, both will flow, the water and the mixture, or rather a mixture which is produced from the two united.

65. A Vessel from which Wine or Water may be made to flow separately or mixed

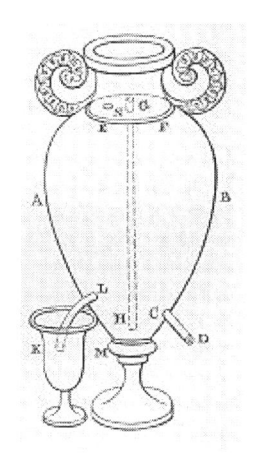

IF water is poured into a vessel standing upon a pedestal and provided with a spout somewhat above its bottom, at one time pure water flows out, at another a mixture of wine and water, and then unmixed wine alone. Let A B (fig. 65), be the vessel, standing upon a pedestal and provided with the spout C D, of which the orifice C is above the bottom of the vessel. Close the neck of the vessel with the partition E F, and through E F insert the tube G H, projecting slightly above the partition and extending to the bottom of the vessel except that a space is left sufficient for the passage of water. In the body of the vessel, and projecting without it, let there be another tube K L, under which a vessel of unmixed wine, K M, is to be placed: in the partition E F pierce a very fine hole N. If, when these arrangements are complete, we pour water into the vessel through the neck, the liquid lying round the projection of the tube will remain in the neck; but all above this will be carried into the body of the vessel, and when it has reached the orifice C of the spout, there will be a discharge of pure water. When a stream has begun to issue from the spout, the unmixed wine in the vessel K M will be drawn up at the same time, on the principle of the siphon, and a mixture will be discharged; and when the water is exhausted, the pure wine will flow by itself, except indeed that the water about the partition E F will be attracted at the same time. When the small quantity of water on E F has all run through N, the air will enter and break the continuity and there will be no further discharge.

66. Wine discharged into a Cup in any required quantity

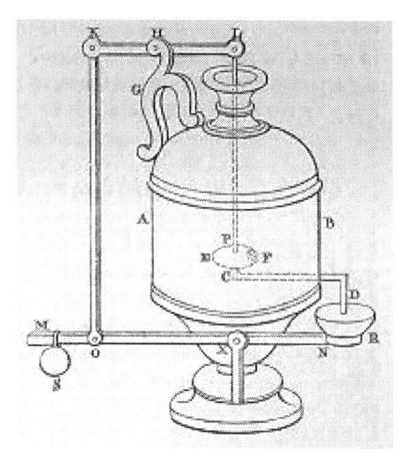

LET there be a vessel filled with wine and provided with a spout under which a drinking cup is placed: wine shall run into the cup in any required quantity. Let A B (fig. 66), be the vessel containing wine, and C D the spout, the upper surface of which at the extremity C is so smooth that, when a valve in the form of a kettle-drum E F is placed upon it, water is excluded. On the handle of the vessel fix the vertical rod G H, on which, as on a fulcrum, another rod K L vibrates: again place another rod, M N, under the pedestal, moving about the point X, and attach two more rods K O, L P, moving on pivots in such a way that, if the extremity M of the bar be depressed, the valve E F is raised, and the spout is opened and sends out a stream, but is closed again when M is suffered to return. Let the bar M N support the drinking-cup R, into which we wish to receive the given quantity of liquid: the cup must be placed beneath the spout. Take a weight, S, capable, by means of a ring, of being shifted along the projection M O of the rod: and when S has been brought towards M, the spout will be opened and send its stream into the cup, but as the cup grows heavy the weight will be raised again and the spout closed. That the wine may flow out in the required quantity, place in the cup any measure of liquid, for instance, a cotyle, and, receiving what falls from the spout in another vessel, shift the weight along the bar to the first point at which the discharge from the spout ceases: make a mark on the bar at this point and register one cotyle. We must proceed in the same manner for a half-cotyle, and two cotyle, and so on for other measures as far as we please; and thus we shall have marks for the different quantities, signifying the points to which the weight must be brought in order that they may be discharged. Instead of the valve E F, an airtight vessel may encircle the spout, so that, as long as the liquid is kept away by the air within, there will be no discharge through the spout.

67. A Goblet into which as much Wine flows as is taken out

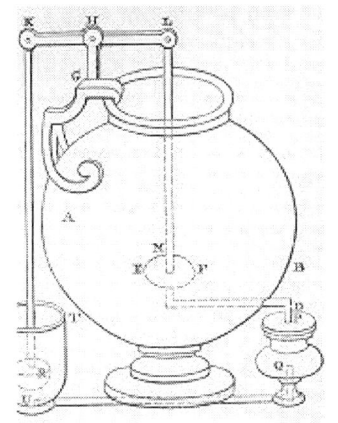

LET there be a vessel containing wine and provided with a spout, underneath which a goblet is placed: whatever quantity of wine is taken from the goblet, as much shall flow into it from the spout. Let A B (fig. 67), be the vessel of wine, and C D the spout, to which are attached the valve E F, and the rods GH, KL, KO, LM as before; and beneath the spout place the cup P. To the rod K O fix a small basin R contained in the vessel S T, and let a tube, U Q, connect the vessels ST and P. When these arrangements are complete, if the vessels S T and P are empty, the basin will fall to the bottom of S T, and open the spout C D. A stream will flow from C D into both the vessels S T and P, so that the basin will rise and shut the spout again, until we remove more liquid from the goblet. This result will happen as often as we remove liquid from P.

68. A Shrine over which a Bird may be made to revolve and sing by Worshippers turning a wheel

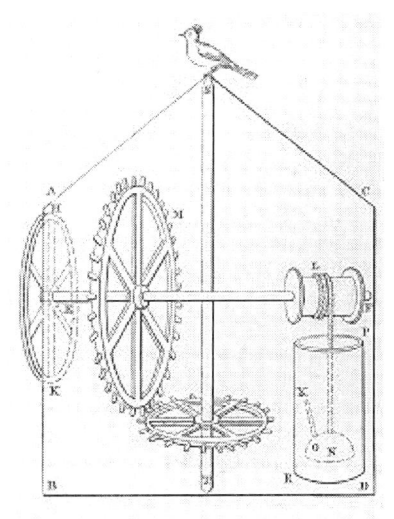

THE construction of a shrine provided with a revolving wheel of bronze, termed a purifier, which worshippers are accustomed to turn round as they enter. Let it be required that, if the wheel is turned, the note of the black-cap shall be produced, and the bird, standing on the top of the shrine, turn round as well; while, if the wheel is turned [in the opposite direction], the black-cap neither sings nor revolves. Let A B C D (fig. 68), be the shrine and E F an axis extending across it, capable of revolving freely, to which the wheel H K, which is to be turned round, is attached. Let two other wheels be attached to the axis, in the interior of the shrine, L and M, of which L has a pulley, and M is a wheel with rays. Round the pulley a cord is wound, from the extremity of which is suspended a vessel N, shaped like a conical oven, and provided with a tube X O, terminating in a small pipe which produces the note of a black-cap: under the conical vessel N must be placed a vessel of water. From the top of the shrine let fall a small axis S T capable of revolving freely: at the extremity S let a black-cap be placed, and at T a wheel with rays, the rays of which are implicated with, or take into, the rays of the wheel M. It will be found that, when the wheel H K is made to revolve, the cord is wound round the pulley and raises the conical vessel N; but, if the wheel is let go, N descends by its own weight into the water and produces the sound by the expulsion of the air. The black-cap turns round at the same time owing to the revolution of the wheels.

69. A Siphon fixed in a vessel from which the Discharge shall cease at will

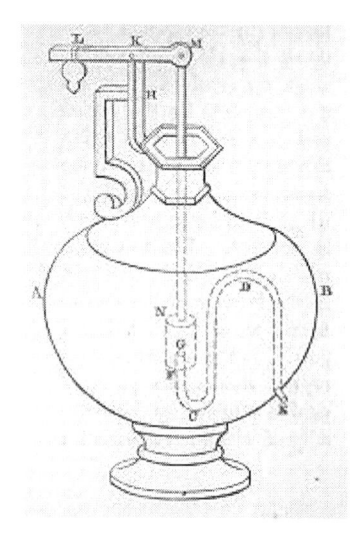

THERE are certain siphons which, when placed in vessels, flow until the vessels are emptied, or the surface of the water has sunk to the level of the outer orifice of the siphon. Let it be required that the discharge shall suddenly cease whenever we wish. A B (fig. 69), is a vessel containing a siphon, C D E, the inner leg of which is bent upwards as at C F G. Let a vertical rod H K be fixed, on which another, L M, works as a lever beam: from L M extends another rod, M N, moving on a pivot, and provided at the extremity N with a vessel large enough to encircle the bent portion of the siphon F G. On the rod L M suspend a weight at L, so that the encircling vessel is raised above the upward bend of the siphon, and the siphon flows. When we wish the discharge to cease, we have only to remove the weight at L, and the vessel at N will descend and encircle the bend G C, so that the siphon will cease to flow. If it is desired that the stream should continue, we must again suspend the weight.

70. Figures made to dance by Fire on an Altar

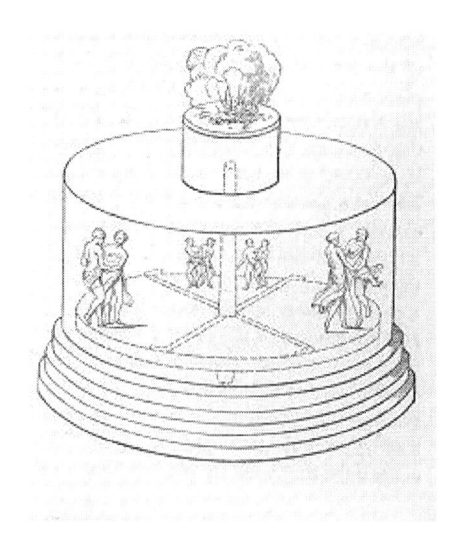

WHEN a fire is kindled on an altar, figures shall be seen to dance: for the altar must be transparent, either of glass or horn. Through the hearth of the altar (fig. 70), a tube is let down turning on a pivot towards the base of the altar, and, above, on a small pipe which is attached to the hearth. Communicating with, and attached to, this tube are smaller tubes lying at right angles to each other, and bent at the extremities in opposite directions. A wheel or platform on which the dancing figures stand, is also fastened to the tube. When the sacrifice is kindled, the air, growing hot, will pass through the pipe into the tube, and be forced out of this into the smaller tubes; when, meeting with resistance from the sides of the altar, it will cause the tube and the dancing figures to revolve.

71. A Lamp in which the Oil can be raised by Water contained within its Stand

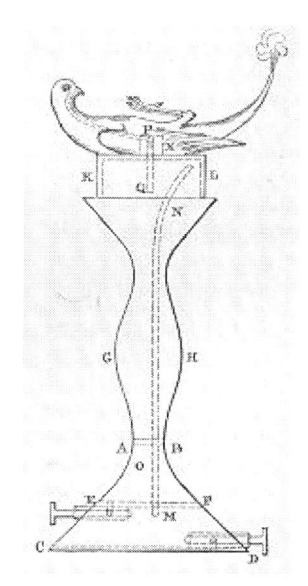

THE construction of a lamp-stand, such that, if a lamp is placed upon it, whenever the oil fails, a supply shall be poured into it from the handle to the amount required, though no vessel is placed upon the lamp from which the oil can flow into it. Let the lamp-stand be constructed with a triangular pedestal, D like a pyramid, A B C D (fig. 71), hollow and provided with a partition E F. Let G H, which must also be hollow, be the shaft of the lamp-stand, and above this shaft place a hollow cup, K L, capable of containing a considerable quantity of oil. From G the partition, E F, and fitting closely into it, a tube, M N, must extend upwards, leaving a passage for the air between its extremity and the covering of the cup, K L, on which the lamp is placed. Through the plate K L insert another small tube X Q a passage being left for water between it and the bottom of the cup: the tube X Q must project a little above the plate K L, and into the projecting part another pipe P is tightly fitted, closed at its upper extremity, and passing through the bottom of the lamp so as to be included within it, that there may be no projection outside. To P solder another pipe, extremely fine, communicating with it, and reaching to the extremity of the handle, so that its stream will be carried into the body of the lamp; this pipe must have an orifice like the others. Under the partition E F let a tap be soldered leading into the chamber C D E F, so that, when it is opened, the water in the chamber A B E F will pass into C D E F. In the plate A B let a fine hole be perforated through which A B E F may be filled with water; the [air] contained in it will pass out through the same hole. We now remove the lamp and fill the cup with oil through the pipe X Q, the air escaping through M N, and again through an open cock in the bottom C D, when any water in C D E F has first flowed out. The lamp having been placed on the top by means of the sliding tube P, when it is required to pour in oil, we must open the tap in the plate E F, and the water in the chamber A B E F passing into C D E F, the air in C D E F will reach the cup through the tube M N, and force out the oil contained in it: the oil will pass into the lamp through the tube X Q and the pipe attached to it. When we wish the oil to stop running, we must shut the cock and the discharge will cease. This process can be repeated whenever it is necessary.

72. A Lamp in which the Oil is raised by blowing Air into it

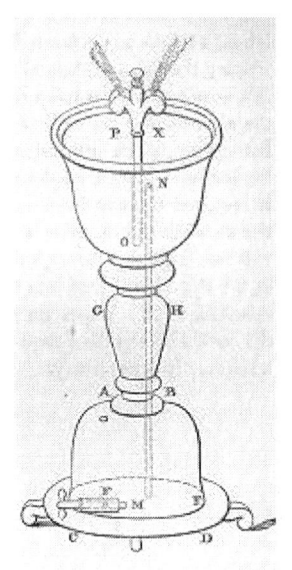

THE same effect can be produced with the same general construction, more readily [than] by constructing the pedestal in which the water is. Let the rest be as before, with the exception of the pedestal and the water in it; the extremity M of the tube M N, (fig. 72), being fitted air-tight into an orifice in the surface of the shaft, so as to be visible outside. Then apply the mouth and blow into the outer orifice; the breath will pass into the cup and force out the oil through the tube X O. Thus the same will take place as before; for as often as we blow into the tube oil will flow into the lamp. It will be necessary that the extremity of the handle should be bent at right angles to the orifice of the lamp, that the oil may not be driven outside.

73. A Lamp in which the Oil is raised by Water, as required

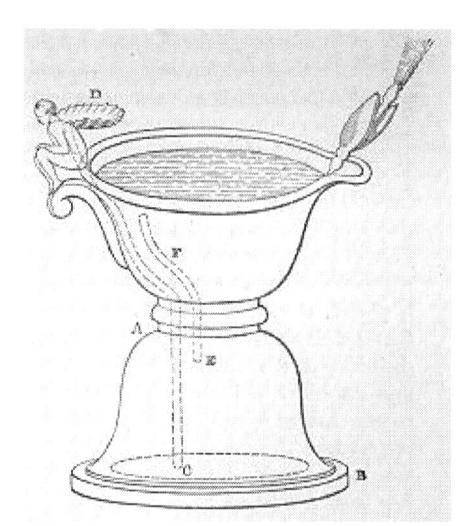

THE construction of a lamp.

Underneath the lamp place a vessel perfectly air-tight, A B (fig. 73), either attached to the lamp or distinct from it. From this let two tubes extend, C D, E F, communicating with the vessel; the extremity C must reach to the bottom except a space sufficient for the passage of water and the tube C D to the surface of the lamp, having at the extremity D a small cup through which the water is to be poured in: the tube E F must pass, air-tight, through the bottom of the lamp. Now if oil be poured through the opening, it will first pass into the vessel A B, and then, when A B is full, the tubes C D, E F, and the lamp will be filled also. As the lamp burns it will become empty, and if we pour in water through the cup, it will pass into the vessel A B, and the oil will ascend and fill up the deficiency in the lamp, until reaches the lamp-nozzle. When the oil has sunk again, we must do the same, repeating it till ⁿly is expended. If it is required to remove the vessel A B, the oil being retained in the ₁ust be a valve or tap in the pipes C D, E F, close to the vessel A B, with keys near ₃o that when the keys are turned, the oil in the lamp, and that in the tubes, shall be confined. Thus the vessel may be removed from the lamp, and, whenever it is desired, we may bring them together again, and open the keys. It is better that the pipe E F should lead to the handle of the lamp, and C D a little behind it, having the cup which communicates with it, and through which water will be poured in, placed above; so that the oil will flow from the handle at the same time that the water is poured into the cup.

74. A Steam-Boiler from which a hot-Air blast, or hot-Air mixed with Steam is blown into the Fire, and from which hot water flows on the introduction of cold

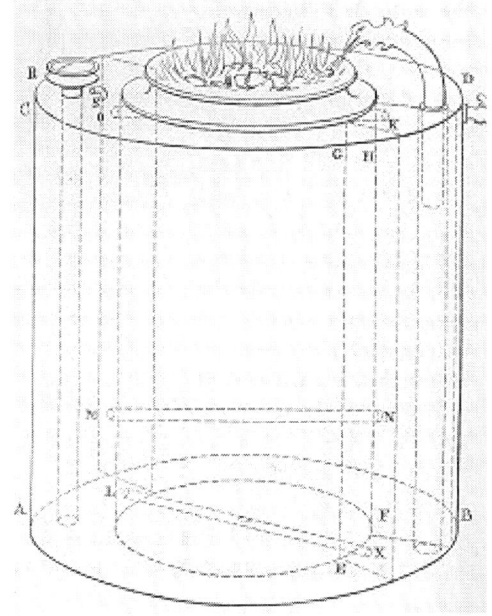

THE construction of a boiler, on which if a figure is placed, shaped as if in the act of blowing, the figure shall blow on the coals and thus the boiler be heated: moreover, if an open spout project near the mouth of the boiler, nothing shall flow from it until we have first poured cold water into a cup; and the cold water shall not mix with the hot until it passes to the bottom of the vessel, while water extremely hot flows from the spout. The shape of the boiler having been determined at pleasure, in that part of it intended to hold the water a small chamber, perfectly air-tight, is intercepted between two perpendicular partitions. With this chamber a tube, one of those which pass under the coals, communicates near the bottom, one end of the tube being closed that no water may enter it from the boiler: the other tubes lead into the chamber where the water is. Thus when the coals are ignited they will generate vapour through that tube which leads into the small chamber. This vapour is carried along a tube which pierces the surface of the boiler, and through the mouth of the figure on to the coals, (for the figure must be bent so as to blow downwards;) and as vapour is always being generated, the figure is always blowing. The vapour is generated from the fire, and, if we pour a very small quantity of water into the small chamber, we shall produce more vapour, and the figure, blowing with great violence, will heat the boiler still higher: just as in the case of cauldrons exposed to fire we see smoke ascending from the water. The figure should be moveable by means of a double sliding tube, to allow of our pouring in the small quantity of water: and, at the same time, by means of this tube, whenever we do not require the figure to blow on the coals, we can turn it round in the opposite direction. On the surface place a small cup from which a tube leads to the bottom of the boiler, that when cold water is poured in, it may pass through to the bottom. In order that the boiler may admit of being filled when water is poured in, and, at the same time, that the water may not boil over and run out, let another pipe communicate with the cup on its inner surface, to avoid offending the sight. We will now expose to view the construction of the boiler. Set up a hollow cylinder (fig. 74), of which A B is the under surface, and C D the upper; and construct another hollow cylinder, with the same axis as the former, of which E F is the under surface, and G H the upper. On the outer edges of the cylinders let plates be fastened, so as to keep the cylinders together and cover the edges. In the cylinder E F G H place the tubes, O K, L X, M N, of which L X perforates the cylinder on one side only at X, while the other two are bored quite through at each end, and their orifices either way open into the space between the cylinders. Into the space intercepted between the two cylinders let down the partitions B G, H F, intercepting the chamber G H E F, into which the tube described above, perforated at one side only, penetrates. Place on the surface, that is on G H, a small tube having the figure attached to, and communicating with, it; the figure must be perforated throughout, and incline downwards so as to look towards the coals. That the figure may cease blowing whenever we like, let the tube on which it sits be fitted tightly into the other, so that, when we turn it round in the opposite direction, the figure will no longer blow on the coals but away from the boiler. We shall also find this sliding tube useful for pouring water into the chamber G F B H, for, after raising the figure from the tube on which it is placed, we can pour the water through, and thus more vapour will be passed along into the figure. On the surface H C let a cup, R S, be placed; communicating with the interior, and having a tube at its extremity reaching down to the bottom of the boiler with the exception of a passage for water. When we desire to let the hot water out, we must pour in cold through R S; this will pass through the tube which communicates with the cup into the chamber of the warm water, which will ascend and flow out through the spout near the neck, for the cold water which has been introduced will not yet have mingled with the warm below. As often as this is repeated we shall obtain warm water for the cold we throw in. In order that we may know when the boiler will bubble up, the chasmatium is contrived, perforated throughout, and placed on the neck, a hole having been made in the surface: it is furnished with a small tube which looks towards the cup R S, that, when the warm water ascends, it may be carried into the cup. Such is the construction of the boiler. If we prefer not to cut off the chamber F G E H through the whole length, but only for a portion of it, the partitions are made to reach half-way, and another is placed upon them, admitting through it a tube which extends up to the figure. When the fire is kindled there will be a rush of vapour from the small chamber, into which water will be poured as before.

75. A Steam-Boiler from which either a hot Blast may be driven into the fire, a Blackbird made to sing, or a Triton to blow a horn

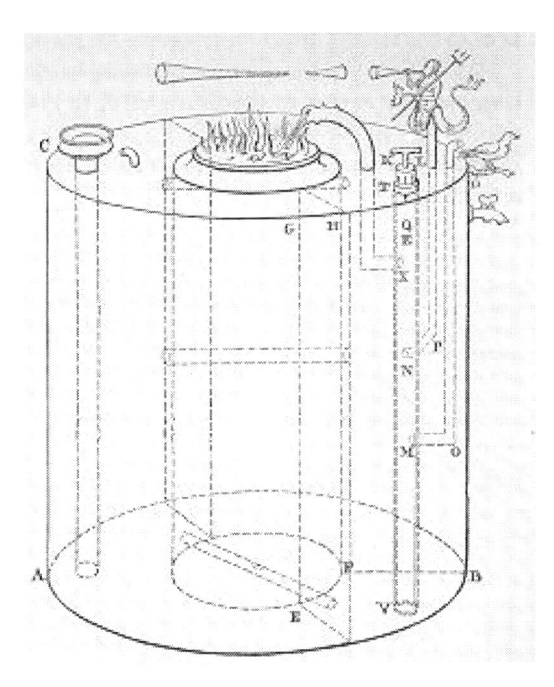

ANOTHER construction of the same kind is employed to produce the sound of a trumpet and the note of a blackbird. A boiler is made (fig. 75), of the same kind as the last, of which all the tubes in the base are bored through at each end, and near the surface there is a tube Q E, into which another tube K L is closely fitted, extending into the chamber for warm air, and moveable about the pin K L. This tube is perforated by three holes, M, N, X, and similarly three holes are bored in Q E opposite the holes M, N, X. Near X, an aperture is made in a support which receives a tube fitting closely into X and surmounted by the figure, as was described in the last paragraph: and from M and N two tubes extend, M O, N P, bent at their upper extremities; these tubes pierce through the surface of the boiler, into which they are carefully soldered. Through the apertures other tubes pass, fitting tightly into the tubes P and O. On one of these tubes is placed the figure of a sparrow, hollow within so as to receive water: the tube on which the bird sits is bent, and provided with a tuning-pipe, such as are made to produce notes, and the curved part of it passes as far as the water contained in the sparrow, so that, when the sound of the pipe reaches the water, the note of a blackbird is produced. In like manner the tube N P has another tube fitting closely into it, on which is placed a figure shaped like a triton with a trumpet in its mouth: the tube on which the triton is placed is moreover furnished with the mouthpiece and bell as usual, and when the vapour reaches these and enters them it will give out the sound of a trumpet. We must discover by trial when the holes in K L are opposite the tubes M O and N P, and when to X on which the figure is placed. Having learnt this, we must make corresponding marks on the pin K L, that the trumpet may sound, or the figure blow, or the blackbird's note be produced, at our pleasure. The arrangements about the cup and the ascent of the warm water are to be made according to the previous description.

76. An Altar Organ blown by manual Labour

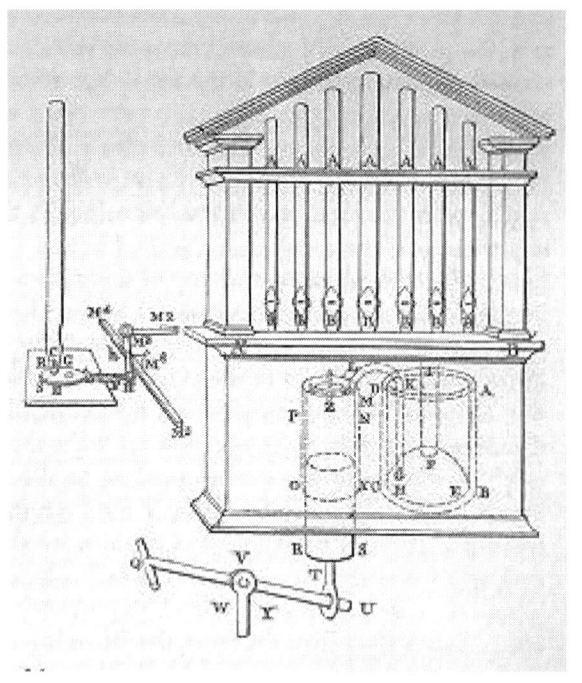

THE construction of a hydraulic organ. Let A B C D (fig. 76), be a small altar of bronze containing water. In the water invert a hollow hemisphere, called a pnigeus, E F G H, which will allow of the passage of the water at the bottom. From the top of this let two tubes ascend above the altar; one of them, G K L M, bent without the altar and communicating with a box, N X O P, inverted, and having its inner surface made perfectly level to fit a piston. Into this box let the piston R S be accurately fitted, that no air may enter by its side; and to the piston attach a rod, T U, of great strength. Again, attach to the piston rod another rod, U Q, moving about a pin at U, and also working like the beam of a lever on the upright rod W Y, which must be well secured. On the inverted bottom of the box N X O P let another smaller box, Z, rest, communicating with N X O P and closed by a lid above: in the lid is a hole through which the air will enter the box. Place a thin plate under the hole in the lid to close it, upheld by means of four pins passing through holes in the plate, and furnished with heads so that the plate cannot fall off: such a plate is called a valve. Again, let another tube, F I, ascend from F G, communicating with a transverse tube, A' B', on which rest the pipes A, A, A, communicating with the tube, and having at the lower extremities small boxes, like those used for money; these boxes communicate with the pipes, and their orifices B, B, B, must be open. Across these orifices let perforated lids slide, so that, when the lids are pushed home, the holes in them coincide with the holes in the pipes, but, when the lids are drawn outwards, the connexion is broken and the pipes are closed. Now, if the transverse beam U Q be depressed at Q, the piston R S will rise and force out the air in the box N X O P; the air will close the aperture in the small box Z by means of the valve described above, and pass along the tube M L K G into the hemisphere: again it will pass out of the hemisphere along the tube F I into the transverse tube A' B', and out of the transverse tube into the pipes, if the apertures in the pipes and in the lids coincide, that is, if the lids, either all, or some of them, have been pushed home.

In order that, when we wish any of the pipes to sound, the corresponding holes may be opened, and closed again when we wish the sound to cease, we may employ the following contrivance. Imagine one of the boxes at the extremities of the pipes, C D, to be isolated, D being its orifice, E the communicating pipe, R S the lid fitted to it, and G the hole in the lid not coinciding with the pipe E. Take three jointed bars F H, H M, M M2, of which the bar F H is attached to the lid S F, while the whole moves about a pin at M3. Now, if we depress, with the hand, the extremity M2 towards D, the orifice of the box, we shall push the lid inwards, and, when it is in, the aperture in it will coincide with that in the tube. That, when we withdraw the hand, the lid may be spontaneously drawn out and close the communication, the following means may be employed. Underneath the boxes let a rod, M4 M5, run, equal and parallel to the tube A' B', and fix to this slips of horn, elastic and curved, of which M6 lying opposite C D, is one. A string, fastened to the extremity of the slip of horn, is carried round the extremity H, so that, when the lid is pushed out, the string is tightened; if, therefore, we depress the extremity M2 and drive the lid inwards, the string will forcibly pull the piece of horn and straighten it, but, when the hand is withdrawn, the horn will return again to its original position and draw away the lid from the orifice, so as to destroy the correspondence between the holes. This contrivance having been applied to the box of each pipe, when we require any of the pipes to sound we must depress the corresponding key with the fingers and when we require any of the sounds to cease, remove the fingers, whereupon the lids will be drawn out and the pipes will cease to sound.

The water is poured into the altar that the superabundant air, (I mean, of course, that which is thrust out of the box and forces the water upwards,) may be confined in the hemisphere, so that the pipes which are free to sound may always have a supply. The piston R S, when raised, drives the air out of the box into the hemisphere, as has been explained; and when depressed, opens the valve in the small box Z. By this means the box is filled with air from without, which the piston, when forced up again, will again drive into the hemisphere. It would be better that the rod T U should move about a pivot at T also, by means of a single [loop,] R, which may be fitted into the bottom of the piston, and through which the pivot must pass, that the piston may not be drawn aside, but rise and fall vertically.

77. An Altar Organ blown by the agency of a Wind-mill

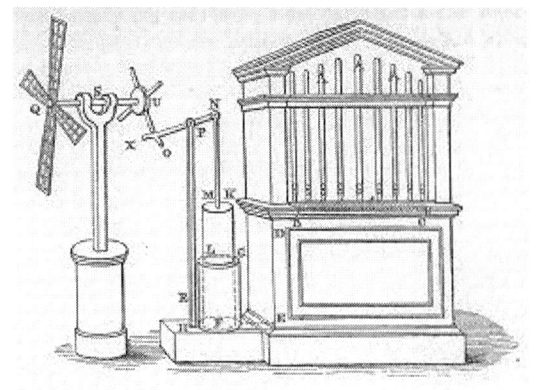

THE construction of an organ from which, when the wind blows, the sound of a flute shall be produced. Let A, A, A, (fig. 77), be the pipes, B C the transverse tube communicating with them, D E the vertical tube, and E F another transverse tube leading from D E into a box G H, the inner surface of which is made level to fit a piston. Into this box fit the piston K L, which is capable of descending into it freely. To the piston attach a rod, M N, and to this another, N X, working on the rod P R. At N let there be a pin moving readily, and to the extremity X fasten a small plate, X O, near which a rod, S, is to be placed, moving on iron pivots placed in a frame which admits of being shifted. To the rod S attach two small wheels, U and Q, of which U is furnished with pegs placed close to the plate X O, and Q with broad arms like the sails of a wind-mill. When all of these arms, urged by the wind, drive round the wheel Q, the rod S will be driven round, so that the wheel U and the pegs attached to it will strike the plate X O at intervals and raise the piston; when the peg recedes, the piston, descending, will force out the air in the box G H into the tubes and pipes, and produce the sound. We may always move the frame which contains the rod S towards the prevailing wind, that the revolution may be more rapid and uniform.

78. An Automaton, the head of which continues attached to the body, after a knife has entered the neck at one side, passed completely through it, and out at the other; the animal will drink immediately after the operation

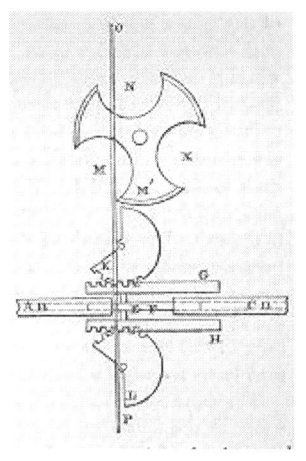

AN animal shall be made to drink while it is being severed in two. In the mouth of the animal (fig. 78), let there be a tube, A B, and in the neck another, C D, passing along through one of the outer feet. Between these tubes let a male cylinder, E F, pass, to which are attached toothed bars, G and H. Above G place a portion of a toothed wheel, K, and, in like manner, beneath H a portion of a toothed wheel, L. Over all let there be a wheel, M', the inner rim of which is thicker than the outer; and let sections be cut out of this wheel by the three circles M, N and X, so that the interval between each division may be equal to the radius of the wheel. Let the rim or felly be likewise divided by the circles, so that the circumference of the wheel will no longer be a circle. Having made an incision, O P, in the upper part of the neck, and severed the head within the incision, make in it a circular Cavity broader below than above, as it were a female tube shaped like an axe, which will contain two sides of the hexagon inscribed in the circle. Let this cavity be S, in which the entire rim M N X will revolve in such a manner that, before one division disappears, the beginning of the next will succeed, and similarly with the third: so that, if a pin be inserted through the wheel, the wheel will revolve, and the head of the animal adhere to the neck. Now, if a knife is passed down through the incision O P, it will enter one of the clefts of the wheel M, and confine it in the circular cavity; and, descending lower, it will touch the projecting tooth of the part K of the wheel, which, being forced downwards, will fit its teeth into those of the bar G, and the bar being pushed back will bring the cylinder out of the tube A B. The knife, passing through the intervening space, will still descend and fall upon the projecting tooth of the part L of the wheel; and this, being forced downwards, and fitting its teeth into the toothed bar H, will drive the cylinder out of C D and fit it into A B. This cylinder is an interior tube fitted into the two tubes, that, namely, in the mouth of the animal, and that reaching from the incision in the neck to the hinder foot. When the knife has passed quite through the neck, and the tube E F has touched both A B and C D, let water be offered to the animal, and a pair of sliding tubes, placed under the herdsman, be turned round. When the herdsman revolves, the water above will flow downwards along the tube C D E F A B, and the current of air caused by the stream of water will attract the water offered to the mouth of the animal. Of course the sliding tubes are so arranged that, as the herdsman turns round, the holes in them coincide. The same result can be brought about without the aid of a stream of water in the following manner. Take once more a pedestal perfectly air-tight, A B C D (fig. 79), having a partition across the middle, E F. Let the tube from the mouth of the animal, G H K, lead into the pedestal, and another tube, L M N, pass through the surface A D and the partition E F. In the tube, L M N, perforate a hole, X, just above the partition E F, and let another tube, O P, fit into it closely, having a hole, R, corresponding with the hole X. To the tube O P attach a figure of Pan, or some other figure with a fierce look, and, when the figure is turned towards the animal, it shall not drink, as though frightened; when the figure turns away, it shall drink. Now, if we pour water into the compartment A D E F through a hole, G', which must afterwards be carefully closed with wax or some other substance, it will be found that, if the holes R and X are made to coincide, the water which was poured in will pass into the compartment E B C F. As A D E F becomes empty, it will attract the air through the mouth of the figure, which will then drink when a cup is presented to it.

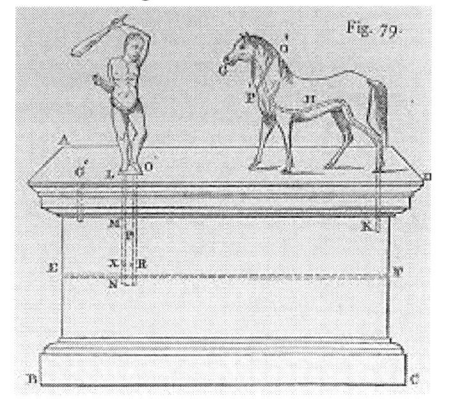

LIBRI LATINI

Puer Zingiberi Panis et Fabulae Alterae

Fabulae Faciles

Insula Thesauraria

De Expugnatione Terrae Sanctae per Saladinum Libellus

Historia Apollonii Regis Tyrii

Historia Regni Henrici Septimi Regis Angliae

Iter Subterraneum

Liber Kalilae et Dimnae

Rebilii Crusonis Annalium

Gesta Romanorum

Fabulae Divales vel Fairy Tales in Latin

ENGLISH BOOKS

Envocation to Priapus
The Life of Apollonius of Tyana
The Pneumatics of Heron of Alexandria
Chinese Sketches
The Broken Road
The Drum
Romance of Lust
Queen Sheba's Ring
The Summons
Nada the Lily

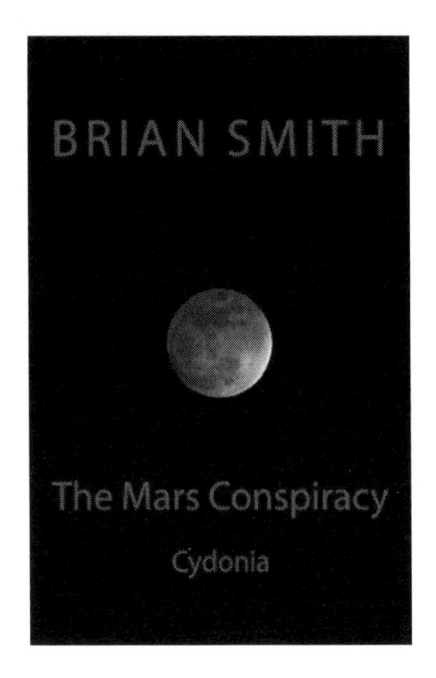

The End of America: & the rest of the world   ISBN-10: 1499534124

Eight gripping apocalyptic short stories about the end of the world. As we all know from Hollywood films the end usually begins in America - the rest of the world is an afterthought. Aliens, doomsday bugs, ancient Greek Gods and even time travel are among the surprising ways that destroy planet Earth. The surprising twists and humour make each of the stories unique and fascinating to read.

The Mars Conspiracy
ISBN-10: 1499620160

Captain Kim Pottinger has been based on Mars for several years and he is feeling bored with the red dustball. All that is about to change when newly arrived Dr. Larry Wathen goes with him to find out why a team of scientists investigating Cydonia, an area famous for the Face and some pyramid shaped mountains, cannot be reached by radio anymore. Of particular concern to Kim is that his girlfriend Jane is one of the missing scientists…

Made in the USA
Middletown, DE
03 June 2023